U.S.NRC

United States Nuclear Regulatory Commission

Protecting People and the Environment

NUREG-1911, Rev. 4

I0482741

NRC Periodic Compliance Monitoring Report for U.S. Department of Energy Non-High-Level Waste Disposal Actions

Annual Report for Calendar Year 2011

Office of Federal and State Materials and
Environmental Management Programs

AVAILABILITY OF REFERENCE MATERIALS
IN NRC PUBLICATIONS

NUREG-1911, Rev. 4

United States Nuclear Regulatory Commission

Protecting People and the Environment

NRC Periodic Compliance Monitoring Report for U.S. Department of Energy Non-High-Level Waste Disposal Actions

Annual Report for Calendar Year 2011

Manuscript Completed: August 2012
Date Published: August 2012

Office of Federal and State Materials and
    Environmental Management Programs

ABSTRACT

This is the U.S. Nuclear Regulatory Commission (NRC) staff's report of its monitoring of U.S. Department of Energy (DOE) non-high-level waste disposal actions in Calendar Year 2011, in accordance with Section 3116(b) of the Ronald W. Reagan National Defense Authorization Act for Fiscal Year 2005 (the NDAA). Section 3116 of the NDAA requires that (1) DOE consult with the NRC on its non-high-level waste determinations and plans, and (2) the NRC, in coordination with the covered States of South Carolina and Idaho, monitor disposal actions that DOE takes to assess compliance with NRC regulations in Title 10 of the *Code of Federal Regulations* (10 CFR) Part 61, "Licensing Requirements for Land Disposal of Radioactive Waste," Subpart C, "Performance Objectives." The NRC has prepared this report in accordance with NUREG-1854, "NRC Staff Guidance for Activities Related to U.S. Department of Energy Waste Determinations," dated August 2007 (NRC, 2007a).

CONTENTS

LIST OF FIGURES

LIST OF TABLES

EXECUTIVE SUMMARY

The purpose of this report is to document the U.S. Nuclear Regulatory Commission (NRC) staff's monitoring of the U.S. Department of Energy (DOE) non-high-level waste disposal actions in Calendar Year (CY) 2011. The NRC monitors DOE disposal actions in covered States in accordance with Section 3116(b) of the Ronald W. Reagan National Defense Authorization Act for Fiscal Year 2005 (NDAA). Section 3116 of the NDAA has two main subsections—subsection (a) requires DOE to consult with the NRC on its non-high-level waste determinations and plans, and subsection (b) requires the NRC, in coordination with the covered States of South Carolina and Idaho, to monitor the disposal actions that DOE takes to assess compliance with NRC regulations in Title 10 of the *Code of Federal Regulations* (10 CFR) Part 61, "Licensing Requirements for Land Disposal of Radioactive Waste," Subpart C, "Performance Objectives." This report is concerned exclusively with subsection (b) of Section 3116. Appendix A to this report provides the complete text of Section 3116 of the NDAA. This is the fourth report of what the NRC anticipates will be an annual report during the early phases of its NDAA monitoring activities. The content of this report follows the guidance in Section 10.4.2 of NUREG-1854, "NRC Staff Guidance for Activities Related to U.S. Department of Energy Waste Determinations," issued August 2007 (NRC, 2007a).

In January 2006, DOE completed the waste determination for salt waste disposal at the Saltstone facility at the Savannah River Site (SRS) in South Carolina (DOE, 2006). DOE issued a second waste determination under Section 3116 on the Tank Farm Facility (TFF) at the Idaho Nuclear Technology and Engineering Center (INTEC) at the Idaho National Laboratory (INL) in November 2006 (DOE-Idaho, 2006). NRC staff (staff) reviewed these waste determinations and associated performance assessments and documented its review in technical evaluation reports (TER) for SRS Saltstone and INL (NRC, 2005a, 2006). In each of these reviews, the staff determined that there was reasonable assurance that DOE's disposal actions would meet the performance objective in 10 CFR Part 61 contingent on the staff's ability to validate certain assumptions through its monitoring role consistent with NDAA Section 3116(b).

Based on the risk-significant monitoring areas identified in these TERs, the NRC completed its initial monitoring plans in for each site (NRC, 2007b, 2007c). In each monitoring plan, the staff identified a hierarchy of elements defining the overall scope of monitoring at each site. The scope of monitoring was defined by those technical subject matter areas identified in the TERs that were most uncertain or significant in the DOE analysis of whether the disposal of these incidental wastes

    i)   **meet the NRC performance objectives and**

    ii)   **can be considered non-high level wastes**

For the Saltstone facility, the NRC staff identified eight risk significant monitoring areas or "factors," which are important model assumptions or parameter values described in its 2005 Saltstone facility TER (NRC, 2005a). For each factor, the NRC has one or more planned monitoring activities (i.e., specific tasks or actions). For the Saltstone facility, NRC staff identified 39 distinct monitoring activities to assess compliance with the performance objectives in 10 CFR Part 61, Subpart C. These 39 monitoring activities are presented in Appendix B, Table B-1.

Similarly, for the INL INTEC TFF, staff identified five risk significant monitoring areas or Key Monitoring Areas (KMAs) from its INL TER (NRC, 2006) that are sub-divided into 31 separate monitoring activities (Appendix B, Table B-2). Monitoring activities can be either onsite observations of disposal activities or technical reviews of documents performed in the office. In this document, the terms "factors" and "KMAs" are both used to refer to technical subject matter areas where staff will focus its monitoring efforts at SRS Saltstone and INL INTEC, respectively. Future revisions of monitoring plans will use consistent terminology across all sites under monitoring to refer to these risk-significant technical subject matter monitoring areas.

In CY 2011, in accordance with the monitoring plans described above, NRC staff continued its technical review of the 2009 Saltstone Performance Assessment (Saltstone PA) and completed two onsite observation visits at the SRS Saltstone facility. There were no active operations at the INL INTEC TFF. Staff did not perform any onsite observations at the site in CY 2011. Staff performed two technical reviews for INL in accordance with monitoring the facility.

In CY 2011, the staff's monitoring activities resulted in no findings of noncompliance, no identification of any new open issues, and no additional recommendations. The staff continued to follow up on two open issues identified in CY 2007 at the Saltstone facility and one open issue identified in CY 2009. The staff has continued to monitor DOE progress on closing open issues in CY 2012. Table 4-1 and Table 4-2 in the body of this report summarize the NRC staff's open issues and recommendations. The body of this report presents more information about the staff's observations, including several follow-up actions that were identified for the Saltstone facility that the NRC and DOE staffs will continue to discuss during CY 2012 observation visits. Section 3.0 presents the two technical reviews completed for INL in CY 2011. Appendix C contains monitoring activity timelines showing monitoring activities from 2007 to 2011 at both SRS Saltstone and INL TFF. Appendix D contains the onsite observation reports for the Saltstone facility.

Savannah River Site Saltstone Facility

In CY 2011, the NRC staff completed two onsite observations to the Saltstone facility (NRC, 2011a; 2011b). In January 2011, DOE provided a tour of Vault 4 and an overview of saltstone production operations in CY 2010. In April 2011, the NRC and DOE staffs discussed the saltstone radionuclide inventory, new research on long-term testing waste oxidation and technetium release, Disposal Unit 2 construction, and summarized the status of 11 issues discussed during previous observations. No new open issues were determined for the Saltstone facility from these observation visits; however, several follow-up actions were identified, which will be discussed in future observation visits. The three previously open issues (Open Issue 2007-1, 2007-2, and 2009-1) were discussed during these observation visits as summarized below.

Open Issue 2007-1 and 2007-2

As discussed above, staff has identified eight monitoring factors that represent risk-significant areas in the DOE analysis of whether the disposal of these incidental wastes meets the NRC performance objectives at the Saltstone facility. The observation of DOE saltstone grout processing and disposal operations is related to Factor 1 ,"Oxidation of Saltstone", and Factor 2, "Hydraulic Isolation of Saltstone," identified in the NRC monitoring plan for the Saltstone facility (NRC, 2007b). The general objectives of NRC monitoring activities related to Factors 1 and 2 are to ensure that the saltstone grout that is produced is of sufficient quality such that there is reasonable assurance that the performance objectives of 10 CFR Part 61 will be met. The hydraulic and chemical properties of the saltstone grout are important for isolating the radioactivity contained in the saltstone grout from the environment (NRC, 2005a, 2005b). A specific objective of the monitoring at the Saltstone facility is to ensure that the saltstone grout formulation produced in the Saltstone Production Facility (SPF) [1] and emplaced in the Saltstone Disposal Facility (SDF) is consistent with the design specifications assumed in the final waste determination (DOE, 2006), or that significant deviations from design specifications will not negatively impact the expected performance of the saltstone grout.

During an observation visit in October 2007, staff observed that DOE had not generated hydraulic and chemical properties of saltstone grout over the range of compositions actually produced at the SPF. The NRC staff concluded in its observation report (NRC 2008a) that additional data over a range of compositions will greatly improve confidence in predictions of future performance of the SDF. The staff also observed that, at the end of a production run, DOE uses water to flush transfer lines between the SPF and SDF. The flush water is added directly to the SDF and may be blending with grout that has not yet set. Staff believes that if the flush water blends with the saltstone grout that has not yet set in the SDF, the water to cement ratio of this portion of the saltstone grout would be much higher than that assumed in the waste determination. Very high water to cement ratios could result in the affected fraction of the saltstone grout having inferior hydraulic properties that could impact the ability of the waste form to meet the performance objectives in 10 CFR Part 61. The staff identified these issues as Open Issues 2007-1 and 2007-2, respectively, in NUREG-1911, "NRC Periodic Compliance Monitoring Report for U.S. Department of Energy Non-High-Level Waste Disposal Actions, Annual Report for Calendar Year 2007," issued August 2008 (NRC, 2008b).

During the April 2011 observation visit, progress on obtaining characterization data for these two open issues was discussed. DOE described plans to continue efforts to determine the hydraulic and chemical properties of as-emplaced saltstone grout. DOE indicated it would complete analysis of existing saltstone core samples and use formed-core sampling to verify the characteristics of as-emplaced saltstone. DOE is developing an integrated sampling plan to correlate the properties of laboratory-prepared and as-emplaced saltstone samples. DOE indicated it was working to quantify variability in the dry feed and the water-to-premix ratios. DOE also indicated it is working to test the hydraulic and physical properties of saltstone formed with various dry feed compositions and cure temperature profiles. As this work is currently ongoing, both Open Issue 2007-1 and Open Issue 2007-2 remain open.

Open Issue 2009-1

The third Open Issue (Open Issue 2009-1) at the Saltstone facility also relates to Factor 1, "Oxidation of Saltstone", with the specific monitoring activity relating to modeling of saltstone oxidation and technetium release. Increased releases of technetium-99 from the waste form could impact compliance with the performance objectives identified in 10 CFR 61.41 and 10 CFR 61.42. In March 2009, staff observed that DOE provided insufficient support for assumptions made regarding the sorption capabilities of the saltstone waste form with respect to K_d values assumed in the 2005 performance assessment (DOE, 2005) and the reduction capabilities of technetium-99 (Tc-99) in the saltstone waste form. To address Open Issue 2009-1, DOE needs to demonstrate that (1) technetium-99 in salt waste is converted to its reduced chemical form in saltstone grout during the curing of saltstone grout, and is thereby

---

[1] This report refers to the "Saltstone facility" which includes both the Saltstone Disposal Facility and the Saltstone Production Facility.

strongly retained in saltstone grout, and (2) the sorption of dissolved technetium-99 onto saltstone grout and vault concrete is consistent with K_d values for technetium-99 that were assumed in the PA.

During CY 2011, NRC and DOE staffs discussed the significant research DOE conducted in this area. DOE measured K_d values up to ~700 mL/g for technetium to saltstone formulated with 45 percent slag (nominal concentration) under a nitrogen atmosphere with 2 percent hydrogen gas. Staff questioned whether results obtained in an atmosphere with 2 percent hydrogen are applicable to as-emplaced saltstone. DOE measured less sorption (K_d of 139 mL/g) of technetium-99 onto cores of saltstone taken from Vault 4, cell E (SRNL-STI-2010-00667) (ADAMS Accession No. ML111310222). DOE hypothesized that the K_d value was significantly less than 1,000 mL/g because 30-60 parts per million oxygen present in the glove box oxidized the saltstone.

Staff conducted independent research with the Center for Nuclear Waste Regulatory Analysis to determine the leachability of several redox sensitive radionuclides including technetium-99, selenium, and uranium. As discussed in "Experimental Study of Contaminant Release from Reducing Grout" (CNWRA and NRC, 2011), low-activity waste was mixed with cementitious grout to create a saltstone waste form. Two types of experiments were conducted with this simulated saltstone to determine the release behavior of the redox-sensitive radioelements technetium, uranium, and selenium initially sequestered in reducing grout as water interacted with the grout and changed the system chemistry. One type of experiment flowed oxygen-bearing simulated SRS ground water through a column of crushed and sieved simulated SRS saltstone material and monitored the changes in pH, E_h, and aqueous concentrations. Technetium release from the simulated saltstone increased sharply during the first 10 pore volumes, increased more gradually until 52 pore volumes in Cell 1 or 26 pore volumes in Cell 2, then afterwards increased significantly with increasing pore volume. The technetium that was released early likely represents technetium that was not effectively immobilized in the reducing grout or technetium that was reoxidized during the crushing and sieving of the grout material. The data also show that uranium is retained in the reducing grout, whereas almost all of the selenium is released after 132 pore volumes.

Based on information DOE provided to NRC during the April 2011 observation, NRC reviewed: (1) DOE experimental efforts to verify that technetium is in fact initially reduced in the saltstone waste form and (2) DOE efforts to provide an estimate of the release rates of oxidized technetium (NRC, 2011b). DOE proposed to close Open Issue 2009-1 based on the results of its recent research, however, the NRC suggested that a complete response to the open issue would indicate whether this range of oxygen concentrations could be present in the as-emplaced saltstone environment. As work continues in this area, Open Issue 2009-1 remains open.

Updated Saltstone PA

In November 2009, staff began its review of the "2009 Performance Assessment for the Saltstone Disposal Facility at the Savannah River Site," (updated Saltstone PA) dated October 2009 (DOE, 2009), and the associated documentation provided. This review was performed in accordance with the NRC's monitoring plan (NRC, 2007b) Section *3.1.9, Performance Assessment Process Review*. Staff continued its review of this PA in CY 2011 and completed its review in CY 2012, as documented in the "Technical Evaluation Report for the Revised Performance Assessment for the Saltstone Disposal Facility at the Savannah River Site, South

Carolina" (NRC, 2012a). A summary of the TER will be included in the annual monitoring report for CY 2012.

Idaho National Laboratory, Idaho Nuclear Technology and Engineering Center, Tank Farm Facility

As mentioned above, staff identified five risk significant KMAs from its INL TER (NRC, 2006). During CY 2011, staff conducted technical reviews in two of these areas, KMA 3 (Hydrogeologic Uncertainty) and KMA 4 (Monitoring during Operations). Staff identified no open issues in CY 2011 for INL INTEC TFF.

KMA 3

KMA 3 was developed as a result of staff's review of the INTEC TFF draft waste determination and supporting PA as documented in NRC (2006), which showed a number of uncertainties associated with DOE's ground water model used to support its demonstration of compliance with the performance objective found in 10 CFR 61.41 for protection of the general population from releases of radioactivity. As stated in the monitoring plan for the INTEC TFF (NRC, 2007c), staff plans to continue to stay abreast of relevant monitoring and modeling activities conducted by DOE, other agencies, or independent researchers until such time that NRC staff can confidently conclude that overall system performance was adequately studied and constrained. If issues related to engineered barrier system performance arose during evaluation of KMA 2, then KMA 3 would become increasingly important. Therefore, staff determined that the status of KMA 3 would remain open until KMA 2 was closed.

Current risks associated with tank farm soil and INTEC ground water from previous releases include external exposure to soil contaminated with cesium-137 and ingestion of contaminated Snake River Plain aquifer (SRPA) ground water. If left unmitigated, perched water could become a continuing source of ground water contamination to the SRPA above certain CERCLA action levels (e.g., maximum contaminant levels or MCLs) beyond 2095. Thus, remedial activities are focused on the control of recharge to the subsurface.

In CY 2011, staff reviewed (1) DOE's annual monitoring report, "Fiscal Year 2010 Annual Operations and Maintenance Report for Operable Unit 3-14, Tank Farm Soil and INTEC Ground water," (DOE-Idaho, 2011b), and (2) DOE's report "Five-Year Review of CERCLA Response Actions at the Idaho National Laboratory Site—Fiscal Years 2005-2009," (DOE- Idaho, 2011a). During the FY 2010 reporting period DOE conducted ground water sampling at 14 SRPA wells and five additional wells sampled as part of the Idaho CERCLA Disposal Facility monitoring program. Data were consistent with previous data revealing the highest technetium-99 concentrations near or southeast of the INTEC Tank Farm. The highest strontium-90 concentrations were also observed in wells southeast of the Tank Farm. All wells show stable or declining trends.

With respect to perched water and ground water, DOE concluded (DOE-Idaho, 2011a) that the CERCLA response actions were functioning as intended and that previous exposure assessment assumptions remain valid. Although remedial activities are not yet complete and their ultimate effectiveness cannot be assessed at this time, DOE concludes that indications are favorable that the desired effect of these remedies will be achieved. NRC staff agrees with this assessment.

During FY 2010, DOE contractors performed a modeling analysis that addressed an NRC staff recommendation made during NRC's 2010 onsite observation (Recommendation 2010-2). NRC staff recommended that DOE consider recent data collected under the CERCLA program that indicate that anthropogenic sources of water associated with INTEC operations, rather than Big Lost River (BLR) seepage, are a more significant source of perched water currently observed at INTEC TFF. NRC staff reviewed DOE's modeling analysis (Portage, 2011) that showed while the doses would increase by roughly a factor of two, performance objectives could still be met.

NRC staff identified no new and significant information that would invalidate NRC staff's TER conclusions. Information on infiltration rates and the mobility of radiological constituents will continue to be assessed by NRC staff through review of INTEC monitoring data and other sources of information. BLR seepage near the INTEC TFF will also continue to be evaluated to determine its potential impact on ground water flow and transport mechanisms near the TFF. NRC staff continues to have reasonable assurance that performance objectives will be met for the INTEC TFF facility.

KMA 4

KMA 4 in the NRC's TER for INTEC TFF addresses DOE compliance with the performance objective found in 10 CFR 61.43 related to protection of individuals during operations. Although various activities, including the demolition of 31 structures previously associated with the grouted tanks occurred at the site, no major closure activities that may impact the dose to workers and members of the public occurred at the INTEC TFF during CY 2010. Dose to workers and member of the public that occurred during CY 2011 will be evaluated in the Annual Monitoring Report for CY 2012.

NRC staff collected and reviewed monitoring data from DOE's 2010 environmental surveillance reports (DOE-Idaho, 2011c), the Idaho DEQ INL Oversight Program annual report for calendar year 2010 (Idaho DEQ, 2011c) and Idaho DEQ's quarterly surveillance reports for the first and second quarters of 2011 (Idaho DEQ, 2011b, 2011c). NRC staff used this information to evaluate the impacts of INL operations on members of the public as well as evaluate the air, soil, water, vegetation, animals, and foodstuffs on and around the INL site to confirm compliance with applicable laws and regulations. Data reported were generally consistent with historic trends. Concentrations of radioactivity in air, soil, and milk samples were consistent with background levels. Radiation levels were also consistent with historic background measurements. All radionuclide concentrations in ambient air samples were below DOE standards and are considered to have no measurable impact on the environment. The maximum dose to the maximally exposed individual was calculated to be well below the applicable radiation protection standard of 0.1 mSv/year (10 mrem/year).

Staff believes that the consistency between data collected by Idaho DEQ and DOE provides confidence that both programs can be used to evaluate offsite environmental impacts associated with INL operations. Based, in part, on the environmental surveillance data collected by DOE and the State, NRC staff continues to have reasonable assurance that the 10 CFR 61.43 performance objective related to protection of individuals during operations will be met.

NRC staff will continue to evaluate worker and public exposure data or estimates through review of worker radiation records and review of environmental surveillance reports as the INTEC TFF closure activities progress in support of the technical review activities identified for KMA 4 in the INL monitoring plan (NRC, 2007c). The level of monitoring is expected to be higher during active closure operations conducted through the year 2012.

Conclusion

Based on its observations and technical review activities for CY 2011, staff concluded that it continues to have reasonable assurance that the applicable criteria of the NDAA can be met if key assumptions made in the DOE waste determinations prove to be correct.[2] In accordance with the requirements of the NDAA and consistent with the NRC's monitoring plans, staff will continue to monitor DOE disposal actions at SRS and INL. The staff expects the monitoring activities to be an iterative process, and several onsite observation visits and technical reviews of various reports, studies, and other documents may be necessary to obtain the information needed to close all of the current open issues, as well as issues that may be opened in the future.

---

[2] Note that staff concluded that it no longer had reasonable assurance that the Saltstone facility could meet the performance objectives in 10 CFR Part 61, "Licensing Requirements for Land Disposal of Radioactive Waste," Subpart C, "Performance Objectives," in the SDF TER issued in April 2012 (NRC, 2012a). This conclusion was reached following completion of staff's CY 2011 monitoring activities documented in this report.

ABBREVIATIONS

ADAMS	Agencywide Documents Access and Management System
ALARA	As Low As Reasonably Achievable
ARP/MCU	actinide removal process and modular caustic side solvent extraction unit
BLR	Big Lost River
CAP88-PC	Clean Air Act Assessment Package 1988
CERCLA	Comprehensive Environmental Response, Compensation, and Liability Act
CFR	Code of Federal Regulations
CY	Calendar Year
DEQ	(Idaho) Department of Environmental Quality
DOE	U.S. Department of Energy
EM	Environmental Monitoring
HLW	High-Level Waste
HRR	Highly Radioactive Radionuclide
I-129	iodine-129
INL	Idaho National Laboratory
INTEC	Idaho Nuclear Technology and Engineering Center
ICRP	International Commission on Radiological Protection
K_d	distribution coefficient
KMA	key monitoring area
LFRG	Low-Level Waste Disposal Facility Federal Review Group
LLW	low-level waste
MDIFF	mesoscale diffusion air dispersion model
MEI	Maximally Exposed Individual
mrem	Millirem
mrem/yr	Millirem per year
µSv	Microsievert
µSv/yr	Microsievert per year
NDAA	Ronald W. Reagan National Defense Authorization Act for Fiscal Year 2005
NRC	U.S. Nuclear Regulatory Commission
PA	Performance Assessment

SC DHEC   South Carolina Department of Health and Environmental Control
SDF   Saltstone Disposal Facility
SPF   Saltstone Production Facility
SRPA   Snake River Plain Aquifer
SRS   Savannah River Site

TER   technical evaluation report
TFF   Tank Farm Facility
Tc-99   technetium-99

WD   waste determination

1.0 PURPOSE OF THIS REPORT

The purpose of this report is to aggregate all monitoring activities performed at each site specified by Section 3116 of the Ronald W. Reagan National Defense Authorization Act for Fiscal Year 2005 (the NDAA). While not required by law, this report is intended to be consistent with NRC's policy on openness. NRC seeks to keep the public informed about U.S. Nuclear Regulatory Commission (NRC) monitoring of the U.S. Department of Energy's (DOE's) radioactive waste disposal process at these sites. NRC also seeks to keep the covered States informed by documenting monitoring activities in coordination with the covered States.

1.1 Background

In October 2004, the U.S. Congress passed legislation that allows the Secretary of Energy to determine, in consultation with the NRC, whether radioactive waste resulting from the reprocessing of spent nuclear fuel is not high-level radioactive waste. The legislation in Section 3116 of the NDAA requires that (1) the DOE consult with the NRC on its non-High-Level Waste (HLW) determinations and plans, and (2) that the NRC, in coordination with the covered State, monitor DOE disposal actions to assess compliance with NRC regulations in Title 10 of the *Code of Federal Regulations* (10 CFR) Part 61, "Licensing Requirements for Land Disposal of Radioactive Waste," Subpart C, "Performance Objectives." The covered States under Section 3116 of the NDAA are South Carolina and Idaho.

Under the NDAA, as part of DOE's consultation with the NRC, DOE will identify specific inventories of radioactive waste and associated facilities and equipment (e.g., tanks, piping, disposal cells) that are candidates for non-HLW decisions. The Secretary's decision is based on whether the residual radioactive waste meets several criteria in Section 3116 of the NDAA. For example, the subject of a Secretary's decision may be residual radioactive waste remaining in an HLW storage tank after the *Highly Radioactive Radionuclides* (HRR) have been removed to the maximum extent practicable. Appendix A to this report provides the full text of Section 3116 of the NDAA, including the criteria.

To support the Secretary's decision, DOE prepares a document, called a *waste determination* (WD), which describes its basis for a determination under Section 3116 of the NDAA. This document describes DOE's analysis of whether a particular type of waste meets the NDAA criteria. In addition to the WD, DOE prepares a *performance assessment* (PA) to predict long-term disposal site performance (see Section 1.3). As described in NUREG-1854, "NRC Staff Guidance for Activities Related to U.S. Department of Energy Waste Determinations," issued August 2007 (NRC, 2007a), staff consults with DOE on the draft waste determination, reviews the assumptions and parameters included in DOE's PA, and prepares a *Technical Evaluation Report* (TER) that documents the NRC staff's evaluation. If the Secretary decides that all of the Section 3116 criteria are met, the Secretary may make a non-HLW determination, and DOE may publish a final waste determination.

After the Secretary's determination, and based on the conclusions in NRC's TER, the NRC staff will, in coordination with the covered State and as described in NUREG-1854 (NRC, 2007a), prepare a written plan to monitor DOE's disposal actions for the purpose of assessing compliance with the *performance objectives* established in 10 CFR Part 61, Subpart C. Table 1-1 presents the performance objectives from 10 CFR Part 61, Subpart C.

Table 1-1: Performance Objectives of Part 61, Subpart C

Section	Title	Text
§61.40[3]	General Requirement	Land disposal facilities must be sited, designed, operated, closed, and controlled after closure so that reasonable assurance exists that exposures to humans are within the limits established in the performance objectives in 10 CFR 61.41 through 10 CFR 61.44.
§61.41[4]	Protection of the General Population from Releases of Radioactivity	Concentrations of radioactive material that may be released to the general environment in ground water, surface water, air, soil, plants, or animals must not result in an annual dose exceeding an equivalent of 25 millirems to the whole body, 75 millirems to the thyroid, and 25 millirems to any other organ of any member of the public. Reasonable effort should be made to maintain releases of radioactivity in effluents to the general environment as low as is reasonably achievable.
§61.42	Protection of Individuals from Inadvertent Intrusion	Design, operation, and closure of the land disposal facility must ensure protection of any individual inadvertently intruding into the disposal site and occupying the site or contacting the waste at any time after active institutional controls over the disposal site are removed.
§61.43	Protection of Individuals during Operations	Operations at the land disposal facility must be conducted in compliance with the standards for radiation protection set out in 10 CFR Part 20 of this chapter, except for releases of radioactivity in effluents from the land disposal facility, which shall be governed by 10 CFR 61.41. Every reasonable effort shall be made to maintain radiation exposures as low as is reasonably achievable.
§61.44	Stability of the Disposal Site after Closure	The disposal facility must be sited, designed, used, operated, and closed to achieve long-term stability of the disposal site and to eliminate to the extent practicable the need for ongoing active maintenance of the disposal site following closure so that only surveillance, monitoring, or minor custodial care are required.

Because NRC monitoring is risk-informed and performance-based, it focuses on assumptions, parameters, and features that are expected to have either a large influence on the performance demonstration or relatively large uncertainties, or both.

As of the end of CY 2011, DOE has completed two waste determinations in consultation with the NRC since the NDAA was enacted in 2004. The first, in January 2006, was the waste determination for salt waste disposal at the Savannah River Site (SRS) in South Carolina

---

[3] In general, to assess compliance with the requirements of 10 CFR 61.40, the NRC will rely on its assessment of DOE's compliance with 10 CFR 61.41 through 10 CFR 61.44. Specifically, the NRC will view DOE as being in compliance with 10 CFR 61.40 as long as DOE is deemed to be in compliance with the other performance objectives.

[4] As stated in the staff requirements memorandum for SECY-05-0073, "Implementation of New USNRC Responsibilities under the National Defense Authorization Act of 2005 in Reviewing Waste Determinations for the U.S. DOE," dated June 30, 2005 (NRC, 2005b), the dose standard is 25 millirem (mrem) total effective dose equivalent using the methodology of the International Commission on Radiological Protection (ICRP), Publication 26, "Recommendations of the International Commission on Radiological Protection" (ICRP,1977).

(DOE, 2006). DOE issued a second waste determination under Section 3116 on the Tank Farm Facility (TFF) at the Idaho Nuclear Technology and Engineering Center (INTEC) in November 2006 (DOE-Idaho, 2006). DOE submitted a draft waste determination for the F-Area Tank Farm facility at SRS in CY 2011, and the NRC issued a TER documenting its review of the draft waste determination in October 2011 (ML112371715). DOE submitted the final waste determination for the F-Area Tank Farm facility to the NRC in CY 2012, therefore, monitoring activities at the F-Area Tank Farm will be included in the CY 2012 version of this annual monitoring report.

Staff prepared a TER (NRC, 2005a, 2006) for each facility that identifies risk-significant parameters and assumptions DOE used in its PA for each site. Based on these TERs, staff developed monitoring plans (NRC, 2007b, 2007c) for each facility. Section 1.2 of this report summarizes the staff's approach to developing monitoring plans for DOE facilities in covered States. Additionally, DOE, on its own initiative, occasionally consults with the NRC on its non-HLW determinations at the Hanford site in the State of Washington and the West Valley Demonstration Project in the State of New York. However, neither Washington nor New York are covered States under the NDAA. Therefore, the NRC does not have a monitoring role at these sites under Section 3116 of the NDAA, and this report does not address these sites.

1.2 The NRC's National Defense Authorization Act Monitoring Approach

Section 10, *NDAA Compliance Monitoring*, of NUREG-1854 (NRC, 2007a) describes, in detail, the staff's approach to compliance monitoring in accordance with Section 3116 of the NDAA. This section summarizes some of the information in Section 10 to provide context for the staff's observations.

Section 3116(b)(1) of the NDAA requires that the NRC shall "in coordination with the covered State, monitor disposal actions taken by the Department of Energy...for the purpose of assessing compliance with the performance objectives set out in Subpart C of Part 61 of Title 10, Code of Federal Regulations." Therefore, as described below, the staff develops its monitoring plans in coordination with the covered States of Idaho and South Carolina.

As mentioned previously, the basis for the monitoring plan for a facility is NRC's TER that documents the review of DOE's WD, PA, and other supporting documents. The NRC has adopted a risk-informed and performance-based approach to monitoring DOE disposal activities under Section 3116 of the NDAA. A cornerstone of the NRC's approach is the identification of *key monitoring areas* (KMAs), or "monitoring factors" related to DOE disposal actions that should be the focus of its monitoring efforts. KMAs are programmatic or technical subject matter areas critical to DOE's ability to demonstrate compliance with the performance objectives of 10 CFR Part 61, Subpart C. The focus of KMAs is generally to build confidence in DOE models and parameters. Staff identifies one or more *monitoring activities* to support each KMA (or monitoring factor) in facility-specific monitoring plans. The performance objectives, KMAs, and monitoring activities form a hierarchy of plan elements that serves as the structure of each monitoring program. The factors and associated monitoring activities identified for the Saltstone facility are listed in Table B-1. The KMAs and associated monitoring activities identified for INL INTEC TFF are listed in Table B-2. In future revisions of NRC monitoring plans, a consistent terminology will be chosen across the DOE sites to designate these technical subject matter areas, either KMAs or factors.

Figure 1-1 illustrates the hierarchy of elements in an NRC monitoring plan by illustrating a hypothetical example of the relationship among 10 CFR Part 61 performance objectives, a single monitoring area, and the different types and categories of monitoring activities. Section 1.3 summarizes the staff's process for developing these elements.

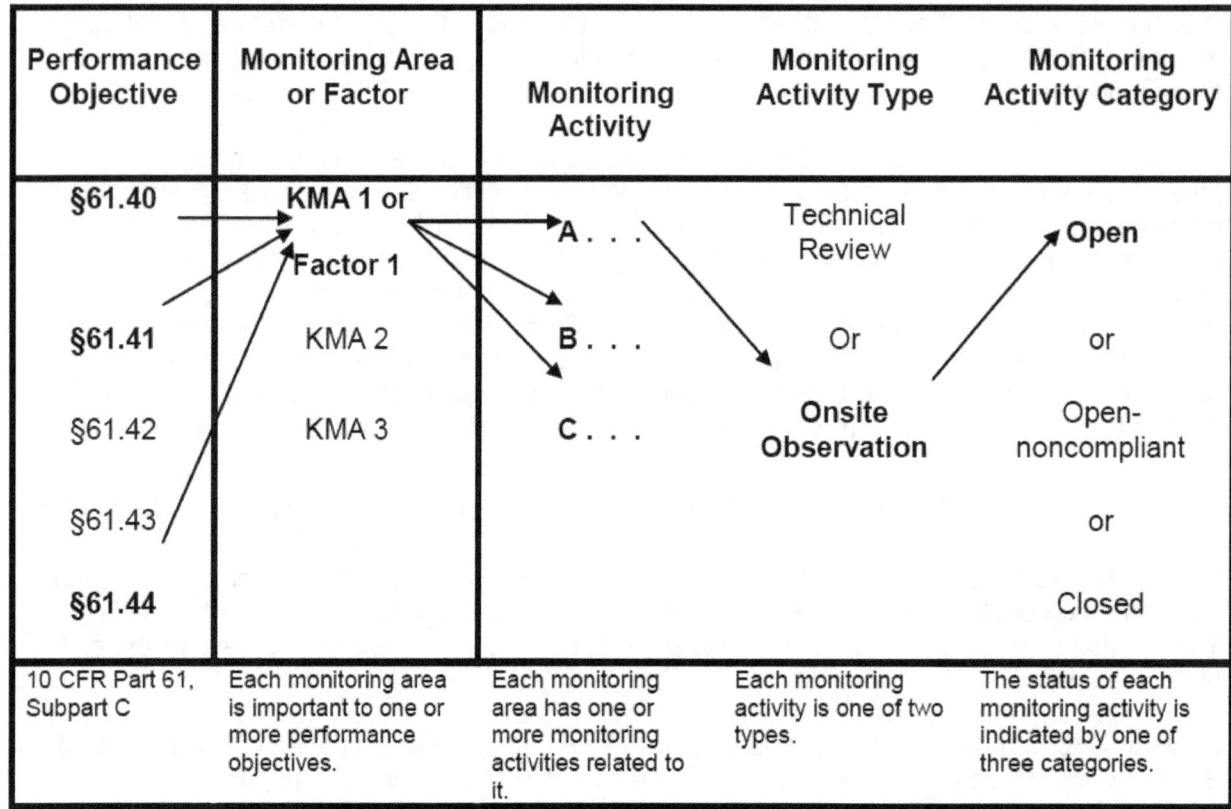

Performance Objective	Monitoring Area or Factor	Monitoring Activity	Monitoring Activity Type	Monitoring Activity Category
§61.40	KMA 1 or Factor 1	A . . .	Technical Review	Open
§61.41	KMA 2	B . . .	Or	or
§61.42	KMA 3	C . . .	Onsite Observation	Open-noncompliant
§61.43				or
§61.44				Closed
10 CFR Part 61, Subpart C	Each monitoring area is important to one or more performance objectives.	Each monitoring area has one or more monitoring activities related to it.	Each monitoring activity is one of two types.	The status of each monitoring activity is indicated by one of three categories.

Figure 1-1: Hypothetical example of relationships between monitoring elements

1.3   Key Monitoring Areas

As the first step in the preparation of a monitoring plan for a specific waste determination, staff identifies the KMAs or monitoring factors. These KMAs focus staff's monitoring efforts in areas that are important to DOE's ability to demonstrate compliance with the performance objectives of 10 CFR Part 61, Subpart C (see Table 1-1). The NRC staff typically identifies the monitoring areas during its review of the DOE draft waste determination, and associated PA, and documents them in the TERs.

Staff determines whether the requirements of 10 CFR 61.41, "Protection of the General Population from Releases of Radioactivity," 10 CFR 61.42, "Protection of Individuals from Inadvertent Intrusion," and 10 CFR 61.44, "Stability of the Disposal Site after Closure," will be met on the basis of DOE predictions of long-term disposal site performance. As described further below, DOE uses a PA to predict disposal site performance, which most often involves calculations performed with the aid of computer-based models. Each site's PA makes certain assumptions about physical and chemical parameter values that DOE believes are appropriate for the disposal action. As such, monitoring areas that build confidence in the DOE selection of parameters and models are typically designated as KMAs.

A PA is an important tool used by both DOE and the NRC to identify which facility attributes are important to meeting the 10 CFR Part 61, Subpart C, performance objectives. In fact, DOE typically uses a PA to demonstrate compliance with the requirements in 10 CFR 61.41, 10 CFR 61.42, and 10 CFR 61.44, recognizing that long-term modeling evaluations are needed to demonstrate compliance with performance objectives. A PA is a type of systematic risk analysis that addresses (i) what can happen, (ii) how likely it is to happen, (iii) what the resulting impacts are, and (iv) how these impacts compare to specifically defined standards. Staff believes that sufficient PA model support, coupled with observation of disposal actions carried out in conformance with detailed closure plans, is necessary for the staff to assess whether these performance objectives can be met in the future. Therefore, the designation of KMAs under 10 CFR 61.41, 10 CFR 61.42, and 10 CFR 61.44 is generally related to the assumptions and parameter values chosen by DOE in its basis documents (i.e., the PA and WD).

Staff identified additional monitoring areas related to 10 CFR 61.43, "Protection of Individuals During Operations." These additional monitoring areas are not typically derived from the staff's review of a DOE PA, as are KMAs. For example, the requirements of 10 CFR 61.43 apply to facility *operations*, including DOE site programs for ongoing personnel site access control, *worker* and public radiation protection, and environmental monitoring (EM) and surveillance. These DOE site programs are required to ensure compliance with the 10 CFR 61.43 performance objective, but are not evaluated as part of the long-term PA of the disposal facility, which as mentioned above is used to demonstrate compliance with 10 CFR 61.41, 10 CFR 61.42, and 10 CFR 10.44.

As noted in Table 1-1, there are generally no specific monitoring areas tied to 10 CFR 61.40, "General Requirements." Staff will rely on its assessment of DOE compliance with 10 CFR 61.41 through 10 CFR 61.44. Specifically, the NRC will view DOE as being in compliance with 10 CFR 61.40 as long as DOE is deemed to be in compliance with the other performance objectives.

1.4 Monitoring Activities

The next step in the preparation of a monitoring plan is the designation of one or more monitoring activities associated with each monitoring area. A monitoring activity is a specific type of NRC or covered State task or action with the purpose of monitoring DOE disposal actions to assess compliance with the performance objectives listed in 10 CFR Part 61, Subpart C. Examples of monitoring activities include staff (NRC and the covered State) reviewing the results of DOE measurements of residual radioactivity in tanks before tank closure, observing periodic maintenance of disposal facility closure caps, and observing onsite radiation safety procedures during waste-handling operations. These examples show that some monitoring activities are near-term, short-duration activities that the NRC or covered States will close soon after the completion of the DOE disposal action. Other monitoring activities are long term, and the NRC or the affected covered State staff may conduct them in perpetuity.

In a few instances, the staff identified monitoring activities during preparation of the monitoring plan that the corresponding TER did not previously identify. As a result, these activities are not related to any particular monitoring area, but are tied directly to a 10 CFR Part 61, Subpart C, performance objective. Examples would include environmental data and performance assessment process (i.e., PA update) reviews.

For staff's planning purposes, monitoring activities are also categorized by type as either technical reviews or onsite observations. Technical reviews may take the form of reviews of data, such as from environmental management and surveillance programs, or reviews of technical literature that supports important assumptions or parameter values in DOE PAs. Data reviews are a subset of, and supplement to, technical reviews that focus on real-time monitoring data that may also indicate future system performance (e.g., sampling and analysis of perched water underneath grouted vaults for changes in chemical conditions) or review of records or reports that can be used to directly assess compliance with performance objectives (e.g., review of radiation records). Onsite observations are coordinated with the affected covered State and the DOE site to ensure that the NRC staff has an opportunity to observe specific DOE disposal actions. The staff conducts onsite observations in accordance with observation plans that are prepared in advance of the visits. The staff summarizes its conclusions in an observation report typically issued within two months of the onsite observation, unless DOE provides additional information following the site visit. In those cases, the reports are typically finished within 60 days of the staff completing its review of the additional information.

Based on their status, staff tracks key monitoring activities as either an *open activity*, an *open-noncompliant activity*, or a *closed activity*. The NRC characterizes a monitoring activity as an open activity when it has not obtained sufficient information to fully assess compliance with one or more 10 CFR Part 61 performance objectives. Should an ongoing open activity provide evidence that the performance objectives of 10 CFR Part 61 are currently not being met, or will not be met in the future, or if key aspects of the waste determination relied on to demonstrate compliance with the performance objectives are no longer supported, then the monitoring activity is categorized as an open-noncompliant activity. The staff's TER and initial monitoring plan may also identify an open-noncompliant activity when the staff finds that the draft waste determination provides insufficient technical bases to determine that the performance objectives will be met. Finally, staff may categorize an ongoing monitoring activity as closed when it has either obtained sufficient information or received technical bases to fully assess compliance with one or more 10 CFR Part 61, Subpart C, performance objectives. However, staff may, upon evaluation of new information, reopen a closed activity or open a new monitoring activity relating to any monitoring area. Any DOE revisions to its PAs may also trigger a review and possible revision of the NRC's monitoring plans.

The following example is provided to illustrate the monitoring process, shown in Figure 1-2. In its 2005 PA for SRS Saltstone, DOE assumed limited oxidation and release of the saltstone waste form based on low diffusion rates of dissolved oxygen into what was assumed to be a relatively low conductivity waste form. Oxidation of saltstone is important to the 10 CFR 61.41 and 10 CFR 61.42 compliance demonstration as it determines the rate at which a key radionuclide, technetium-99, is released from the disposal facility.

In its 2006 TER, staff expressed concerns regarding DOE PA assumptions related to the rate of saltstone oxidation based on a number of factors including the assumed hydraulic properties and degradation of the saltstone waste form over time. Staff developed three monitoring factors (see Section 2.3.4) related to this technical issue including: (1) oxidation of saltstone, (2) hydraulic isolation of saltstone, and (3) model support. NRC issued a monitoring plan for the Saltstone facility in May 2007 that provided specific monitoring activities related to these factors, as summarized in Table B-1. In 2009, staff created an "open issue" to track this concern. Open Issue 2009-1 specifically states that DOE needs to demonstrate that (1) technetium-99 in salt waste is converted to its reduced chemical form in saltstone grout during the curing of saltstone grout, and is thereby strongly retained in saltstone grout, and (2) the sorption of dissolved technetium-99 onto saltstone grout and vault concrete is consistent with K_d values for technetium-99 that were assumed in the PA.

During CY 2011, NRC and DOE staffs discussed this Open Issue extensively during the April 2011 observation visit (see Section 2.3.1.2.2). Based on information DOE provided to NRC during this observation, NRC reviewed: (1) DOE experimental efforts to verify that technetium is in fact initially reduced in the saltstone waste form and (2) DOE efforts to provide an estimate of the release rates of oxidized technetium (NRC, 2011b). These technical review activities support "Factor 1—Oxidation of Saltstone" and its associated monitoring activity "Review field and laboratory experiments and any additional modeling of saltstone oxidation and technetium release" (SRS-SLT-41-01-03-T) and "Factor 3 – Model Support" and its associated monitoring activity "Review DOE conceptual model for oxidation and technetium release and any support for the model" (SRS-SLT-41-03-04-T) listed in Table B-1.

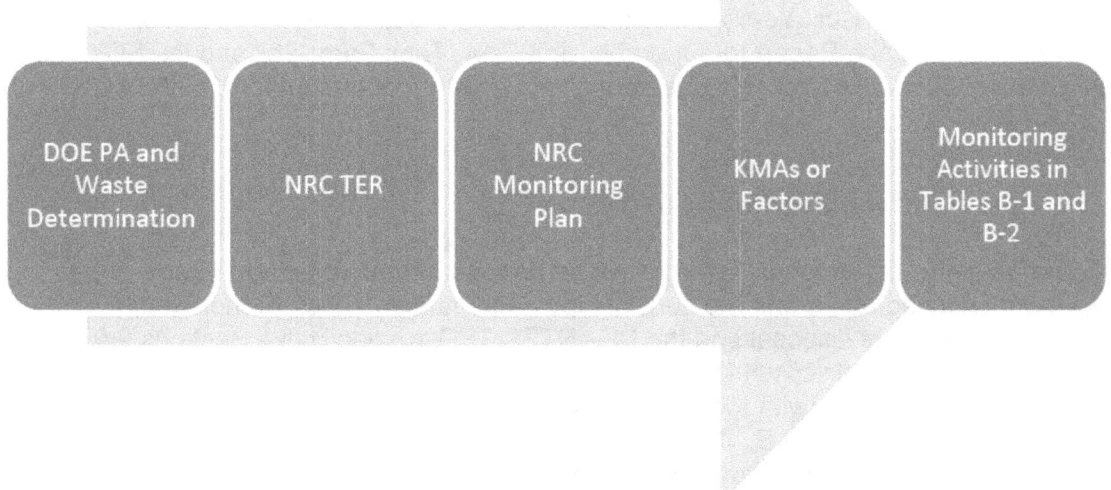

Figure 1-2: Diagram of Monitoring Process

1.5   Coordination with Covered States

Staff consulted with the States of South Carolina and Idaho during the preparation of the monitoring plans for Saltstone and the Idaho National Laboratory (INL) INTEC TFF. For Saltstone, the staff had early interactions with the South Carolina Department of Health and Environmental Control (SC DHEC) during its review of the waste determination and later sought comments on the draft monitoring plan. As a result of these interactions, the staff considered the regulatory activities of South Carolina relating to both a State wastewater permit for the Saltstone Production Facility (SPF) and a State industrial solid waste permit for the Saltstone Disposal Facility (SDF) in the development of its plan. Because of the combined roles of SC DHEC and the NRC under Section 3116(b), the staff operates in a manner to leverage South Carolina's activities pertaining to these permits and avoid duplication of effort.

In CY 2011, staff coordinated each onsite monitoring activity at the Saltstone facility with the State of South Carolina. At least one state representative was present onsite during the

January 2011 observation visit. No SC DHEC personnel were available to attend the April 2011 observation visit; however, the results of the observation were communicated to the State.

Similarly, for the INL INTEC TFF, the staff engaged the Idaho Department of Environmental Quality (DEQ) early in the consultation process during the staff's review of the DOE waste determination. The two primary State regulatory responsibilities related to the TFF are (1) Resource Conservation and Recovery Act closure under the Hazardous Waste Management Act, and (2) Comprehensive Environmental Response, Compensation, and Liability Act (CERCLA) regulatory activities associated with historical releases from the ancillary equipment associated with the TFF that resulted in soil and ground water contamination. In its monitoring plan and in practice, the NRC considered these and other non-regulatory environmental surveillance activities and has leveraged Idaho's activities to avoid duplication of effort. For example, the NRC routinely relies on site reports published by Idaho DEQ for independent surveillance. As it does every year, staff reviewed DOE's environmental surveillance reports and Idaho DEQ's quarterly surveillance reports for the first and second quarters of 2011 (DOE-Idaho, 2011c; Idaho DEQ, 2011a; and Idaho DEQ, 2011b), as discussed in Section 3.3.2. Staff also reviewed DOE-Idaho's "Five-Year Review of CERCLA Response Actions at the Idaho National Laboratory Site—Fiscal Years 2005-2009" and the "Fiscal Year 2010 Annual Operations and Maintenance Report for Operable Unit 3-14, Tank Farm Soil and INTEC Ground water" (DOE, Idaho 2011a, 2011b). No observation visits were conducted for INL INTEC TFF in CY 2011.

1.6   Status of Monitoring Activities

Table B-1 and Table B-2 in Appendix B to this report use the format depicted in Table 1-1 to summarize the monitoring areas and the current types and categorization of monitoring activities for SRS salt waste disposal and the INL INTEC TFF, respectively. Sections 2.0 and 3.0 in the body of this report discuss the monitoring activities in detail for each site. Monitoring plans developed in consultation with the covered States (NRC, 2007b, 2007c) provided the information presented in Appendix B. Timelines for the various monitoring activities conducted from 2007 to 2011 at each site are presented in Appendix C.

2.0 MONITORING AT THE SAVANNAH RIVER SITE SALTSTONE FACILITY IN CALENDAR YEAR 2011

2.1   Introduction

As noted in Section 10.1, *Overall Approach and Scope* of the NRC staff guidance document, NUREG-1854 (NRC, 2007a), the staff's approach to assessing compliance with the performance objectives consists of two primary activities: (1) conducting technical reviews of DOE data and analyses, and (2) physically observing DOE's disposal actions through onsite visits. Since monitoring activities began at the Saltstone facility in 2007, the NRC has completed 11 onsite observations, 11 formal technical reviews, and various data reviews. Each monitoring activity is associated with a public document describing the details of the activity. Each onsite observation is preceded by an onsite observation guidance document, which states the objectives of the observation and the relationship between each objective and its respective 10 CFR Part 61 performance objective. Following the observation, staff documents the activities that took place during the observation in an onsite observation report, which provides an assessment of the staff's activities while on the observation, how those activities relate to their respective 10 CFR Part 61 performance objective, and what conclusions were made from the observations activities.

2.2   Background

On March 31, 2005, DOE submitted the "Draft Basis for Section 3116 Determination Salt Waste Disposal at the Savannah River Site" to demonstrate compliance with the Section 3116 criteria, including demonstration of compliance with the performance objectives in 10 CFR Part 61 (DOE, 2005). In its consultation role, staff reviewed the draft waste determination and concluded that there was reasonable assurance that the applicable criteria of Section 3116 could be met, provided certain assumptions made in DOE's analyses are verified via monitoring. The NRC documented the results of its review in a TER issued in December 2005 (NRC, 2005a). DOE issued a final waste determination in January 2006, taking into consideration the assumptions, conclusions, and recommendations documented in the NRC's TER (DOE, 2006).

On May 3, 2007, the NRC completed its monitoring plan for the Saltstone facility in accordance with the guidance in NUREG-1854 (NRC, 2007a). The monitoring plan covers DOE disposal actions at the Saltstone facility at the Savannah River Site (SRS) in South Carolina. The staff identified a hierarchy of elements defining the overall scope of monitoring at the site. The scope of monitoring was defined by those factors that were most uncertain or significant in the DOE analysis of whether the disposal of non-high-level waste meets NRC performance objectives, which are aimed at the protection of public health and safety. Staff identified eight "factors" that are important model assumptions or parameter values described in its December 2005 TER (NRC, 2005a). For each *factor*, the agency has one or more planned monitoring activities (i.e., specific tasks or actions). For Saltstone, 39 distinct monitoring activities exist to assess compliance with the performance objectives in 10 CFR Part 61. These monitoring activities are presented in Table B-1 in Appendix B. Monitoring activities can be either onsite observations of disposal activities or in-office reviews of documents.

To carry out its monitoring responsibility under NDAA, the NRC performs three types of activities: (i) technical reviews, (ii) onsite observations, and (iii) data reviews in coordination with the State of South Carolina site regulator, SC DHEC. These activities focus on the eight factors mentioned above, and are also identified in the NRC monitoring plan for salt waste disposal at SRS (NRC, 2007b). Technical reviews are generally focused on reviewing additional model support for assumptions DOE made in its PA that are considered important to DOE's compliance demonstration. Onsite observations generally are performed to (i) observe the collection of data (e.g., observation of waste sampling used to generate radionuclide inventory data) and review the data to assess consistency with assumptions made in the waste determination, or (ii) observe key disposal (or closure) activities related to technical review areas (e.g., slag and other material storage, grout formulation and preparation, and grout placements). Data reviews supplement technical reviews by focusing on monitoring data that may indicate future system performance or by reviewing records or reports that can be used to directly assess compliance with performance objectives.

As the staff completes technical reviews and onsite observations, it may identify *open issues* that arise during monitoring activities that require additional follow-up by the staff or additional information from DOE to address questions the NRC staff has raised regarding DOE disposal actions. Since inception of NRC monitoring of the Saltstone facility in 2007, the NRC has identified four open issues and has closed one of these issues (NRC, 2008b). A summary of these open issues can be found in Section 4.0 of this report.

The following terms are used to classify the topics discussed in the Section 2.3.

Remains Open: The NRC is still awaiting action on the part of DOE, or results from a recent action taken by DOE. Further discussion will need to take place before the NRC can close the topic.

Topic Closed: The specific inquiry posed by the NRC has been fully responded to by DOE.

Future Consideration: The specific inquiry posed by the NRC has been discussed and DOE has stated a path forward that seems acceptable to the NRC. The item is not open because the DOE plans to address the topic. The item is not closed because the NRC is interested in the results of the analysis being performed by DOE.

Recommendations may address: (1) ways in which DOE can make progress on closing any open activities in the staff's monitoring plan, (2) a monitoring area for which an open issue has been previously identified and closed and for which staff recommends further action to strengthen some aspect of the DOE disposal action, or (3) monitoring areas that had no open issues or previously raised concerns, but for which staff recommends further improvements in DOE disposal actions.

Appendix C provides a visual depiction of the timeline of NRC monitoring of the Saltstone facility under NDAA from 2007 to 2011.

2.3   NRC Monitoring Activities in 2011

The NRC staff continued its review of the 2009 Saltstone PA in CY 2011 and completed its review in CY 2012, as documented in the "Technical Evaluation Report for the Revised

Performance Assessment for the Saltstone Disposal Facility at the Savannah River Site, South Carolina" (NRC, 2012a).

The NRC staff also completed two onsite observations at the Saltstone facility in CY 2011. In January 2011, DOE provided a tour of Vault 4 and an overview of saltstone production operations in CY 2010 (NRC, 2011a). In April 2011, the NRC and DOE staffs discussed the saltstone radionuclide inventory, new research on long-term testing waste oxidation and technetium release, Disposal Unit 2 construction, and summarized the status of 11 issues discussed during previous observations (NRC, 2011b). The body of this report presents more information about the staff's observations. Details of each of these observations can be found in Appendix D of this report.

In CY 2011, the staff's monitoring activities resulted in no findings of *noncompliance*. The staff continued to follow up on the two open issues identified in CY 2007 and one open issue identified in CY 2009. The staff has continued to monitor DOE progress on closing open issues in CY 2011. As discussed below, after summarizing the status of the 11 issues discussed during previous observations, several follow-up actions were identified.

2.3.1 Onsite Observations

As reported in the annual monitoring report for CY 2010 (NRC, 2012b), three onsite observations were conducted in February 2010, April 2010, and July 2010. Many of the topics covered during the CY 2010 observation visits were also discussed in CY 2011, when the staff conducted two observation visits: January 27, 2011, and April 26, 2011.

The staff's January 27, 2011, onsite observation at SRS Saltstone was focused on assessing compliance with the four performance objectives: (i) protection of the general population from releases of radioactivity (10 CFR 61.41), (ii) protection of individuals from inadvertent intrusion (10 CFR 61.42), (iii) protection of individuals during operations (10 CFR 61.43), and (iv) stability of the disposal site after closure (10 CFR 61.44), by observing Vault 4 integrity and discussing saltstone production operations. Meeting these performance objectives is predicated on the performance of the disposal cells within the period of compliance. Appendix D to this report contains the observation report dated March 15, 2011.

The staff's April 26, 2011, onsite observation at SRS Saltstone was also focused on assessing compliance with the four performance objectives above. To accomplish these goals, staff discussed testing of saltstone properties, Vault 4 inventory, disposal unit construction, and recent research on technetium-reduction and oxidation in saltstone performed by SRS. Appendix D to this report contains the observation report dated August 19, 2011. Details of these observations are discussed below.

2.3.1.1 January 2011 Onsite Observation

As discussed more fully in the observation report in Appendix D, the observation began with a short briefing presented by the DOE contractor, Savannah River Remediation (SRR) and attended by representatives from DOE, the NRC, SC DHEC, and SRR. The briefing consisted of going through the observation agenda and reviewing standard safety considerations at the facility in preparation of a facility tour. After the briefing, Saltstone Production Facility (SPF) staff (employees of SRR) took the group on a tour of Vault 4, which consisted of observing the exterior wall of an empty vault cell: Cell H. SRR staff then moved the group into a conference room to discuss the operations at SPF and to watch a short video of a seepage spot on Cell F. The individual monitoring areas are listed as subsections below, along with the results for each area.

2.3.1.1.1   Vault 4 Integrity

The observation of DOE saltstone disposal operations pertains to Factors 1 and 2 identified in the NRC monitoring plan for the SRS SPF and SDF (NRC, 2007b), and summarized in Section 2.3.4 of this report. Section 3.1.3, *Hydraulic Isolation of Saltstone*, of the May 2007 monitoring plan (NRC, 2007b) provides the basis for the staff's intended review areas.

The concrete vaults of the SDF are assumed to provide secondary containment for saltstone as well as limit waste form exposure to aggressive chemical conditions. The objectives of this portion of the observation visit were to observe Vault 4 walls, with respect to waste form isolation and stability in the local environment, as well as gain an understanding of the process SRR uses to identify seepage spots on the cell wall and conduct subsequent mitigative actions. Verifying the integrity of the Vault 4 walls is important to assessing the vaults ability to maintain hydraulic isolation of the saltstone waste form which relates directly to ensuring compliance with 10 CFR 61.41. Previously, during the July 2010 observation visit (NRC, 2010b), DOE provided a tour of the interior and exterior of Disposal Cell 2 to provide a visual status of corrective actions taken since leaks were found during the hydro-test in April 2010. During this July 2010 visit, staff noted that if leakage occurred around the bolts used to fasten the drainage system to the vault floors in the new vaults, the existing vaults (1 and 4) may also experience similar leakage.

Results

The staff observed the exterior wall of the Vault 4, Cell H, and noted that seepage had occurred at imperfections in the vault walls as liquid builds up in the gap between the saltstone and vault wall. DOE has applied sealant coatings, a rain shield, certified huts, and a drip pan on the exterior of the vault cells to reduce seepage of liquid to the environment. SRR staff also discussed the use of disposal pads to mitigate the releases. SRR staff stated the plan is to dispose of the pads in Vault 1 or E Area. Toxicity characteristic leaching procedure tests are conducted on the pads but a radionuclide-specific characterization of the pads has not been conducted. Characterization of the radionuclides that are released may provide insight into the stability of the saltstone waste form (e.g., whether or not technetium-99 is retained in the waste form).

The vaults are intended to provide secondary containment for the radioactive saltstone waste form. It is not clear that the flow through the walls of Vault 4, as modeled and assumed in the 2009 Saltstone PA (DOE, 2009) is consistent with observations of seepage. The NRC and DOE staffs agreed to further address this issue in an upcoming observation visit.

NRC staff inquired about the integrity of the roofs of the Vault 4 cells as this provides a degree of hydraulic isolation to the waste form. SRR staff indicated that there are active efforts to reduce the infiltration of rainwater into the cells. NRC staff requested any documentation of repair work to the roofs of Vault 4 to ensure that the assumptions in the 2009 Saltstone PA regarding the hydraulic properties of the roof are consistent with ongoing observations. DOE supplied images of the repair work to the roof of Cell A, Vault 4 (ADAMS Accession No. ML110620217).

2.3.1.1.2   Saltstone Production Facility Operation

The staff's interest in discussing operations at the SPF is to ensure that the production of saltstone grout at the SDF is consistent with the assumptions made in the 2009 Saltstone PA (DOE, 2009). Verifying the suitability of the saltstone production process is also important to assessing the site's radiation protection program which relates directly to ensuring compliance with 10 CFR 61.43. Section 5.2.1, "Radiation Protection Program," of the May 2007 monitoring plan (NRC, 2007b) provides the basis for the staff's intended review areas.

Results

Staff was not able to observe the saltstone grout in operation during the observation. In lieu of observing active operations, SRR staff provided a presentation explaining the current inventory being disposed of onsite, a short description of 2010 operational parameters, and an assessment of any atypical operational parameters (e.g., unusual work stoppages, abnormal worker exposure).

In response to the NRC's request, DOE provided a chart with the details of saltstone production during CY 2010 (ADAMS Accession No. ML110620205). The staff learned that approximately 2,630 kiloliters (694,000 gallons) and 1,481 Terabecquerels (40 kilocuries) of salt solution were disposed of in 2010 and that Vault 4 is expected to be at capacity sometime in early 2012. DOE indicated it was planning to dispose of 2 million gallons during 2011 and "several hundred thousand" gallons more in the beginning of 2012.

SRR staff stated that they believe that the disposal of thorium-230 at the SDF is significantly below the assumed activity in the 2009 Saltstone PA because of a very conservative estimation of thorium-230 inventory in the PA. NRC staff inquired whether the predicted disposal of technetium-99 into Vault 4 is consistent with the assumed activity in the 2009 Saltstone PA. This topic was discussed further during the April 2011 observation.

2.3.1.2   April 2011 Onsite Observation

The April 2011 observation began with a short briefing on the observation agenda and site safety procedures presented by the DOE contractor, SRR, and attended by representatives from DOE, the NRC, Savannah River National Laboratory (SRNL), and SRR. The observation continued with a discussion between NRC, DOE, and associated DOE contractor staff regarding the inventory of Vault 4, the new technetium oxidation research, disposal Unit 2 construction, and various follow-up discussions from previous observations. The individual monitoring areas are listed as subsections below, along with the results for each area.

2.3.1.2.1   Technical Discussion - Saltstone Radionuclide Inventory

As noted in Section 3.1.1.1, "Data Reviews – Radioactive Inventory," of the May 2007 monitoring plan, it is important for staff to verify the radioactive inventory disposed of at the Saltstone Disposal Facility because the inventory is an important factor in the compliance with the performance objective identified in 10 CFR 61.41 and 10 CFR 61.42.

Results

Three main areas were discussed, as further detailed in the observation report in Appendix D. The majority of the discussion focused on the calculated inventory of iodine-129 (I-129) and the inventory of I-129 assumed for Vault 4 in the 2009 Saltstone PA (DOE, 2009).

1.   NRC staff asked DOE to provide the inventory of each radionuclide disposed of in Vault 4 since March 2009.

      DOE provided staff with the document X-CLC-Z-00034, "Inventory Determination of PODD/SA Radionuclides in the Saltstone Disposal Facility Through 9/30/10" (ADAMS Accession No. ML111310276), which provided the requested information. DOE indicated that it would be providing the NRC with the final inventory on an annual basis under monitoring.

2.   NRC staff asked DOE to provide the method used to estimate the predicted thorium-230 in the 2009 PA and the method currently being used to track the inventory of thorium-230 disposed of in Vault 4.

      This issue was discussed in detail during an NRC/DOE public meeting on April 27, 2011. It was not resolved during the observation, but will be addressed by DOE in its response to the NRC staff's second RAI, RAI-2009-02 (NRC, 2010c).

3.   NRC staff asked DOE to indicate how the current inventory in Vault 4 compares to the assumed inventory in the 2009 PA, specifically noting that, based on quarterly monitoring reports, the inventory of I-129 disposed of in Vault 4 appears to exceed that predicted in the 2009 Saltstone PA.

      DOE indicated the I-129 inventory in Vault 4 does not exceed the inventory predicted in the revised PA (SRR-CWDA-2011-00070) (ADAMS Accession No. ML111310182) based on a reevaluation of the inventory of I-129 disposed of to date in Vault 4. DOE noted that the preliminary concentrations of I-129 reported in the quarterly reports were based on estimates determined using the Tank 50 material balance and were not based directly on sample results. DOE performed a recalculation of the I-129 inventory based on the sample results and estimated that the inventory in Vault 4 was 0.16 Curies, compared to the inventory of I-129 in the 2009 PA (0.28 Curies).

2.3.1.2.2   Technical Discussion – New Research on Long-Term Testing Waste Oxidation and Technetium Release

As summarized in Section 2.3.4, "Factor 1 – Oxidation of Saltstone," saltstone oxidation is considered to be important primarily because oxidation can lead to increased releases of technetium-99 from the waste form, which may impact compliance with the performance objectives identified in 10 CFR 61.41 and 10 CFR 61.42.

To address Open Issue 2009-1, DOE needs to demonstrate that (1) technetium-99 in salt waste is converted to its reduced chemical form in saltstone grout during the curing of saltstone grout, and is thereby strongly retained in saltstone grout, and (2) the sorption of dissolved technetium-99 onto saltstone grout and vault concrete is consistent with K_d values for technetium-99 that were assumed in the PA.

Results

DOE discussed its recent research (SRNL-STI-2010-00667 and SRNL-STI-2010-00668) (ADAMS Accession Nos. ML111310222 and ML111310234) with NRC staff during this observation. DOE measured K_d values up to ~700 mL/g for technetium to saltstone formulated with 45 percent slag (nominal concentration) under a nitrogen atmosphere with 2 percent hydrogen gas. NRC staff questioned whether results obtained in an atmosphere with 2 percent hydrogen are applicable to as-emplaced saltstone. In addition, the slag-free control samples had similar measured K_d values for technetium-99, which indicates that the reduction and sorption of the technetium was not caused by the slag and might have been caused by the hydrogen gas instead. DOE indicated that, because the E_h of the leachate decreased with increasing slag concentrations, they conclude that slag controlled the E_h in the reducing cementitious materials.

DOE measured less sorption (K_d of 139 mL/g) of technetium-99 onto cores of saltstone taken from Vault 4, cell E (SRNL-STI-2010-00667) (ADAMS Accession No. ML111310222). DOE hypothesized that the K_d value was significantly less than 1,000 mL/g because 30-60 parts per million oxygen present in the glove box oxidized the saltstone.

For greater detail of the discussion that took place during this part of the observation, please refer to the DOE document provided to the NRC during the observation, SRR-CWDA-2011-00071 (ADAMS Accession No. ML111310199). Based on the results of this recent research, DOE proposed to close Open Issue 2009-1, related to the initial chemical reduction of and the K_d value for technetium-99 in saltstone. The NRC suggested that a complete response to the open issue would indicate whether this range of oxygen concentrations could be present in the as-emplaced saltstone environment.

The NRC conducted independent research with the Center for Nuclear Waste Regulatory Analysis to determine the leachability of several redox sensitive radionuclides including technetium-99, selenium, and uranium. As discussed in "Experimental Study of Contaminant Release from Reducing Grout" (CNWRA and NRC, 2011), low-activity waste was mixed with cementitious grout to create a saltstone waste form. Two types of experiments were conducted with this simulated saltstone to determine the release behavior of the redox-sensitive radioelements technetium, uranium, and selenium initially sequestered in reducing grout as water interacted with the grout and changed the system chemistry. One type of experiment flowed oxygen-bearing simulated SRS ground water through a column of crushed and sieved simulated SRS saltstone material and monitored the changes in pH, E_h, and aqueous concentrations. Technetium release from the simulated saltstone increased sharply during the first 10 pore volumes, increased more gradually until 52 pore volumes in Cell 1 or 26 pore volumes in Cell 2, then afterwards increased significantly with increasing pore volume. The technetium that was released early likely represents technetium that was not effectively immobilized in the reducing grout or technetium that was reoxidized during the crushing and sieving of the grout material. The data also show that uranium is retained in the reducing grout, whereas almost all of the selenium is released after 132 pore volumes.

The second type of experiment leached cylindrical specimens of the simulated saltstone material—one set cured at room temperature and another at 60 degrees Celsius (C) (140 degrees Fahrenheit)—in deionized water and monitored aqueous concentrations over time. Qualitatively, the data indicate that the leach rates for the different species increase in the order technetium < nitrate ≈ nitrite < selenium. Measured uranium concentrations were mostly below the reporting limit, indicating uranium was not released from the reducing grout within the timeframe of the experiment. However, because the physical and chemical conditions in actual saltstone waste and the release behavior of radionuclides could be different than those in the laboratory experiments, this report strongly recommends leaching experiments using actual SRS saltstone samples.

2.3.1.2.3   Discussion of Disposal 2 Unit Construction

The staff's interest in discussing construction activities of the new disposal cells relates to ensuring the integrity of the disposal units and identifying the potential mechanisms of contaminant release from the facility. Section 3.1.3, "Hydraulic Isolation of Saltstone," of the May 2007 monitoring plan (NRC, 2007b) provides details of the basis for the staff's intended review areas.

Results

DOE discussed cell design changes to deal with hydraulic leaks including flush cutting anchor bolts, cold capping of type V concrete without anchor bolt, washer and nut mechanical seals, and flexible coatings (see also Hydrotest Results in section 2.3.1.2.4). DOE provided SRR-CWDA-2011-00082 (ADAMS Accession Nos. ML111320032 and ML111320049), which describe the design changes made to the new disposal cells.

2.3.1.2.4   Follow-up Discussion – Topics from Previous Observations

The staff's interest in discussing the list of topics in this section relates to multiple sections of the May 2007 monitoring plan and also relates to all four of the 10 CFR Part 61 performance objectives.

Results

During the observation, DOE provided a document, SRR-CWDA-2011-00043 (ADAMS Accession No. ML111310214), which contains many of the details of the discussion provided in this section of the report. This section compiles many topics discussed in previous observations and provides the current status of each topic.

1.   Performance Assessment/Research Activity

DOE discussed current and future SDF PA maintenance activities. The current research activities include 11 studies on parameters such as the reducing environment, dispersion coefficients, degradation mechanisms, closure cap infiltration, hydrology, and geology. The planned PA maintenance activities include degradation studies, impacts of waste oxidation, vault cracking and attendant transport, and code upgrade.

Staff would like to know the results of these current research activities. This is not a follow-up action; however, the staff maintains an interest in PA maintenance activities and will continue discussions with the DOE leading up to its upcoming revision to the 2007 NRC monitoring plan for the Saltstone facility.

2.   Open Issues 2007-1 and 2007-2

The observation of DOE saltstone grout processing and disposal operations is related to Factor 1 ,"Oxidation of Saltstone", and Factor 2, "Hydraulic Isolation of Saltstone," identified in the NRC monitoring plan for the Saltstone facility (NRC, 2007b). The general objectives of NRC monitoring activities related to Factors 1 and 2 are to ensure that the saltstone grout that is produced is of sufficient quality such that there is reasonable assurance that the performance objectives of 10 CFR Part 61 will be met. As discussed in the NRC TER for the Saltstone facility, the hydraulic and chemical properties of the saltstone grout are important for isolating the radioactivity contained in the saltstone grout from the environment (NRC, 2005a). A specific objective of the monitoring at the Saltstone facility is to ensure that the saltstone grout formulation produced in the Saltstone Production Facility (SPF) and emplaced in the Saltstone Disposal Facility (SDF) is consistent with the design specifications assumed in the final waste determination (DOE, 2006), or that significant deviations from design specifications will not negatively impact the expected performance of the saltstone grout.

During an observation visit in October 2007, staff observed that DOE had not generated hydraulic and chemical properties of saltstone grout over the range of compositions actually produced at the SPF. The NRC staff concluded in its observation report (NRC 2008a) that additional data over a range of compositions will greatly improve confidence in predictions of future performance of the SDF. The staff also observed that, at the end of a production run, DOE uses water to flush transfer lines between the SPF and SDF. The flush water is added directly to the SDF and may be blending with grout that has not yet set. Staff believes that if the flush water blends with the saltstone grout that has not yet set in the SDF, the water to cement ratio of this portion of the saltstone grout would be much higher than that assumed in the waste determination. Very high water to cement ratios could result in the affected fraction of the saltstone grout having inferior hydraulic properties that could impact the ability of the waste form to meet the performance objectives in 10 CFR Part 61. The staff identified these issues as Open Issues 2007-1 and 2007-2, respectively, in NUREG-1911, "NRC Periodic Compliance Monitoring Report for U.S. Department of Energy Non-High-Level Waste Disposal Actions, Annual Report for Calendar Year 2007," issued August 2008 (NRC, 2008b).

To address Open Issue 2007-1 and 2007-2, DOE should determine the hydraulic and chemical properties of as-emplaced saltstone grout. In addition, DOE should demonstrate that intra-batch variability, flush water additions to freshly poured saltstone grout at the end of each production run, and additives used to ensure processability are not adversely affecting the hydraulic and chemical properties of the final saltstone grout. DOE should show that the hydraulic and chemical properties are consistent with the assumptions in the waste determination or show that any deviations are not significant with respect to demonstrating compliance with performance objectives (NRC, 2008b).

During this observation, DOE described plans to continue efforts to determine the hydraulic and chemical properties of as-emplaced saltstone grout. DOE indicated it would complete analysis of existing saltstone core samples and use formed-core sampling to verify the characteristics of as-emplaced saltstone. DOE is developing an integrated sampling plan to correlate the properties of laboratory-prepared and as-emplaced saltstone samples. DOE indicated it was working to quantify variability in the dry feed and the water-to premix ratios. DOE also indicated it is working to test the hydraulic and physical properties of saltstone formed with various dry feed compositions and cure temperature profiles. Determining the impact of these variations on the performance assessment is planned future work. Staff indicated that the plans to address the open issues sound reasonable. These two issues remain open at this time.

3.   *Open Issue 2009-1*

The discussion on Open Issue 2009-1 is regarding the leachability of technetium-99 and is described in detail in Section 2.3.1.2.2 above. As noted above, this issue remains open at this time.

4.   *Follow-up Action: Disposal Unit 2 Water Tightness Test Quality Assurance Records*

DOE will provide NRC staff with documentation of cell design changes and hydrotesting results for review when they are available following the Operational Readiness Review.
This follow-up action remains open.

5.   *Follow-up Action: Radiological Composition of Inadvertent Transfer Material*

During the July 2010 onsite observation, staff requested information on the radionuclide composition of the salt solution that was inadvertently transferred to Vault 4. DOE provided document SRR-WSE-2010-00186 (ADAMS Accession No. ML111780337) to the NRC on October 26, 2010, in response to this request.

The inadvertent transfer of approximately 7,192 liters (1,900 gallons) of liquid salt solution to Vault 4 occurred on May 19, 2010. This inadvertent transfer was caused by valve misalignment during tests of the salt feed tank agitator. Following the inadvertent transfer, drain water removal was performed to remove the salt waste, and DOE estimates that less than 189 liters (50 gallons) of this material remained on top of the saltstone monolith following this removal. When the SPF was restarted, clean grout was added for 15 to 20 minutes to attempt to encapsulate this remaining liquid. Staff believes it is useful to understand the radiological content of this material because the inventory in the inadvertent transfer material (although limited in quantity) may not be encapsulated in the grout well because it was not disposed of in the form of grout.

The material in the inadvertent transfer consisted of salt waste that originated in Tank 50, plus clean cap drain water returns. A dip sample was taken of the salt waste solution remaining in the hopper at the time of the inadvertent transfer. This sample was characterized for chemical constituents, but the radiological constituents were not characterized (DOE, 2010). DOE estimated the radiological content of the material in the inadvertent transfer based on the radiological composition of the waste from Tank 50 and the estimated dilution from the clean cap drain water.

Staff has reviewed this information and has concluded that while it would have been preferable to have actual radiological characterization data for the material in the inadvertent transfer, the approach used by DOE to estimate the radiological content of this material is reasonable. Based on the information provided to the NRC in SRR-WSE-2010-00186 (ADAMS Accession No. ML111780337), staff considers this action item to be closed.

6.   *Follow-up Action: Status of ARP/MCU Management Control Plan*

The NRC and DOE staffs have discussed the management control plan for the actinide removal process and modular caustic side solvent extraction unit (ARP/MCU management control plan) during the March 2009 onsite observation (ADAMS Accession No. ML091320439), during the April 2011 observation, and in a subsequent phone call on June 30, 2011. Staff stated that the basis for its interest in the status of the ARP/MCU management control plan was that it believed that the sample results obtained under this plan were used for determining the inventory of material transferred from ARP/MCU to Tank 50. Staff was also interested in knowing when the operations under the ARP/MCU management control plan are ceased because DOE had previously indicated samples would be taken less frequently once this happened.

Through these discussions, DOE contractor staff stated that the purpose of the ARP/MCU samples is to obtain information related to safety (such as criticality) and process information and that these samples are not used to develop inventory information for the SDF. Instead, the inventory information is based on direct analytical measurements of the Tank 50 samples and the materials balance calculations for Tank 50.

The inventory assumed for the ARP/MCU feed stream in the materials balance is based on the expected characterization of the waste in the particular salt batch. DOE contractor staff stated that as the actual Tank 50 sample data is made available, the inventory is updated to reflect the sample data, rather than the material balance information.

Because the ARP/MCU sample data did not affect the inventory determination for the SDF, staff considers this follow-up action to be *closed*. However, staff requests that it be informed when any major changes to the salt waste processes are made, such as exiting the ARP/MCU management control plan, as these types of changes will affect the NRC's monitoring activities.

DOE offered to provide a demonstration of the spreadsheet used for these inventory-updating calculations during the next onsite observation. This is not a follow-up action; however, the NRC would like to observe this demonstration in the future. The NRC makes note that this will be a future observation activity.

Additionally, DOE raised the concern that tracking long-term items (such as the exit strategy for the ARP/MCU management control plan) as follow-up actions, might not be the most efficient mechanism. Staff stated that a revised monitoring plan will be developed following the completion of the TER for the revised PA. In the new monitoring plan, staff will generate a list of major changes to the salt waste disposal process that they would like to be made aware of, if and when they occur. The transition from the ARP/MCU management control plan is an example of what would be included on this list, and the follow-up action for DOE to notify the NRC when the management control plan is exited can be handled in this manner in the future. DOE stated that they have no immediate plans to cease operating under ARP/MCU management control plan.

7. *Follow-up Action: Anchor Bolt Penetrations in Vault 4*

During the July 2010 onsite observation, the NRC and DOE staffs discussed leakage from the vault caused by anchor bolts on the floors of cells 2A and 2B during the hydrotests (no waste involved). Staff raised a concern with the integrity of Vault 1 and 4 floors based on the presence of a similar drain system and anchor bolts. Staff suggested that direct evidence of leakage could be determined by horizontal soil cores under Vaults 1 and 4.

During this April 2011 observation, DOE contractor staff updated NRC staff that DOE has visually inspected several anchor bolts locations in cells B and H of Vault 4, which were empty, and did not see any evidence of cracking on the vault floor surface. DOE also discussed historical, semiannual monitoring well data that does not indicate that there have been releases from Vault 4. The potential effects of bolt penetrations will be mitigated in the Future Disposal Cells and the need for bolts (to anchor the cable brace) will be eliminated in the design for the new units. This follow-up action remains open pending the response to the above concern and the completion of the work DOE is performing on this follow-up action. Staff will continue to review documentation regarding this follow-up action as it becomes available.

8.   *Follow-up Action: Impact of Scale on Core Sampling Methodology*

The staff's interest in discussing core sample analysis and sampling procedures relates to ensuring the integrity of the waste form and verifying that the actual saltstone waste form has properties that are consistent with the simulated saltstone samples.

During the July 2010 onsite observation, the NRC and DOE staffs discussed proposed future saltstone core sampling techniques. Alternate methods were discussed and each had its strengths and weaknesses. DOE presented information on an in-situ sampling technique essentially using embedded pipes, for which they tested the force required to remove the sampling device. The NRC expressed concern that the sampling device may allow less disruption of the sample; however, the sampling device may change the in-situ conditions of the waste form such that the sample is not representative. The NRC stated that when its contractor conducted experiments to test the properties of large-scale samples, scale effects were evident in the results (CNWRA, SWRI, and NRC, 2011). This highlights the importance of measuring properties of representative samples at appropriate scale.

During this observation, DOE commented it is now developing a formed-core sampling methodology to minimize the disruption to core samples that was discussed in the July 2010 onsite observation. Staff has commented that formed-core samples may not be representative of in-situ conditions, but it will continue to review core-sampling approaches and results.

DOE plans to move forward with formed-core sampling technology. Operational considerations such as worker exposure and logistics will be considered in the sampling plan. This follow-up action remains open pending the response to the completion of the work DOE is performing on the impact of scale on core sampling methodology.

9.   *Cure Temperatures and Impact of Aluminate Concentration*

An item originally discussed during the October 2007 observation, staff inquired about the cure temperatures for saltstone grout as recent research has indicated its potential significance on the hydraulic properties of saltstone (WSRC-STI-2009-00419). A hydraulic conductivity of 8.6E-7 cm/s was measured for a saltstone grout simulant that was cured at 60 degrees C, which is greater than the value assumed in the Saltstone PA by more than a factor of 400. During the April 2011 observation, DOE stated that cure temperature profiles for saltstone are being compiled and will be considered in future testing. Staff discussed the importance of mimicking field conditions when practicable, including cure temperature and humidity. Staff will review the cure temperature profiles for saltstone when DOE compiles them following future testing.

10.   *Saltstone Fracturing*

During the March 2009 observation, participants watched a video survey that showed fractures on the surface of the saltstone grout in Cell G of Vault 4 (NRC, 2009). The survey area was limited and DOE mentioned during the April 2011 observation that it has since developed a video surveillance program to further evaluate fracturing of the saltstone surface. The video will be analyzed by DOE and evaluated with respect to the PA. Staff will review the video and analysis as they become available.

11.   *Hydrotest Results*

The staff's interest in observing construction relates to ensuring the integrity of the disposal units and identifying the potential mechanisms of contaminant release from the facility. As discussed during the April 2010 observation, the NRC inquired about the details of the recent hydrotests results for each cell (e.g., hydraulic head, test duration, observation procedures). The following year, during the April 2011 observation, DOE stated that the hydrostatic testing of the new disposal cells, following the design changes showed no evidence of leaking. The follow-up testing consisted of a modified version of the earlier hydrotest. The new test consisted of a 12-foot head differential for 132 hours. Staff considers this item to be closed.

2.3.2   Summary of Open Issues, Follow-up Actions, and Recommendations

At the close of CY 2011, the three issues previously identified by the staff remained open: (1) the hydraulic and chemical properties of the saltstone grout, (2) the variability of saltstone from batch to batch, and (3) the reduction and retention of technetium-99 within the saltstone waste form. Further onsite observation visits and technical reviews may be necessary to obtain the information needed to close all of the current open issues, as well as other issues that may be opened in the future. There are no new open issues resulting from the observations conducted in CY 2011. However, there are multiple follow-up actions that were identified during the April 2011 observation as summarized below.

2.3.2.1   January 27, 2011 Observation

No issues or concerns were identified during the observation of Vault 4. With respect to the Vault 4 seepage, the corrective actions taken by DOE should be effective at significantly reducing or eliminating contamination from the vault from reaching the environment in the short term (NRC, 2008c). Staff requested any documentation regarding repair work to the roofs of all Vault 4 cells and maintains an interest in the disposal and characterization of the absorbent pads.

No issues or concerns were identified during the observation of the Saltstone Production Facility operations. NRC and DOE staffs discussed the thorium-230 activity disposed of in Vault 4 to date and an updated prediction of the technetium-99 disposal activity for Vault 4.

2.3.2.2   April 26, 2011 Observation

Saltstone Inventory

Staff believes that the method used in DOE's reevaluation of the inventory of I-129 in Vault 4 seems reasonable and this issue was resolved during the observation. Based on this reevaluation, the current inventory of I-129 in Vault 4 is estimated to be less than the inventory assumed in the 2009 Saltstone PA. The DOE indicated that they would be providing the NRC with the final inventory on an annual basis under monitoring. No additional issues or concerns were identified during the technical discussion on the radionuclide inventory of Vault 4 apart from DOE's continued effort to respond to RAI-2009-02.

New Research – Technetium-99

Because of the staff's concerns about the leachability of technetium-99, Open Issue 2009-1 remains open. No additional issues or concerns were identified during the technical discussion regarding technetium release and oxidation in the saltstone waste form.

Disposal Unit 2 Construction

Staff will continue to monitor the construction of the new disposal cells and will continue to monitor the cells when they are put into operation.

Follow-up Discussion Topics from Previous Observations

The status of the remaining follow-up actions is summarized in the following table. Each topic is classified as being open, closed, or a future topic for discussion.

Discussion Topic	Remains Open	Topic Closed	Future Consideration
PA/Research Activity			X
Open Issues 2007-1 and 2007-2	X		
Open Issue 2009-1	X		
Follow-up Action: Disposal Unit 2 Water Tightness Test Quality Assurance Records	X		
Follow-up Action: Radiological Composition of Inadvertent Transfer Material		X	
Follow-up Action: Status of ARP/MCU Management Control Plan		X	X
Follow-up Action: Anchor Bolt Penetrations in Vault 4/Vault 4 Floor	X		
Follow-up Action: Impact of Scale on Core Sampling Methodology			X
Cure Temperatures and Impact of Aluminate Concentration			X
Saltstone Fracturing			X
Hydrotest Results		X	

2.3.3   Summary of Technical Reviews

A summary of the technical reviews NRC staff completed in CY 2011, including its review of the 2009 Saltstone PA, can be found in the "Technical Evaluation Report for the Revised Performance Assessment for the Saltstone Disposal Facility at the Savannah River Site, South Carolina" (NRC, 2012a). A summary of the TER will also be included in the annual monitoring report for CY 2012.

2.3.4　　　Key Monitoring Factors

2.3.4.1　　　Purpose of Key Monitoring Factors

Staff has identified specific technical areas that will be important for monitoring to assess compliance with the performance objectives during its review of DOE's draft waste determination. The NRC's technical reviews describe key assumptions DOE made in its analyses supporting its salt waste determination and the resulting technical areas, called "factors," that staff plan to monitor to assess compliance with the performance objectives. Staff identified the following eight key factors to monitor: (i) oxidation of saltstone, (ii) hydraulic isolation of saltstone, (iii) model support, (iv) erosion control design, (v) infiltration barrier performance, (vi) feed tank sampling, (vii) Tank 48 waste form, and (viii) removal efficiencies. As mentioned previously, the term "factors" used to track specific technical areas to monitor for the Saltstone facility are analogous to the KMAs that are identified for INL INTEC TFF.

In general, the factors relate to three important aspects of the disposal system: waste form and vault degradation, the effectiveness of infiltration and erosion controls, and estimation of the radiological inventory. Each factor is described in more detail in the sections below.

2.3.4.2　　　Factor 1 - Oxidation of Saltstone

The NRC based its assessment of compliance for the performance objectives on a 10,000-year performance period. Because of the long performance period, several of the monitoring factors relate to the long-term degradation of saltstone and the concrete vaults that the saltstone will be poured into. Chemical oxidation of saltstone was identified as a monitoring factor primarily because of the possibility of unacceptable technetium doses if saltstone is oxidized more rapidly than DOE predicts. To confirm DOE's assumptions about saltstone oxidation, NRC staff expects to monitor the development of better predictions of saltstone oxidation during the 10,000-year performance period and the resulting release of technetium. Specifically, staff expects to monitor the results of oxidation experiments and refined radionuclide release models, among other possible activities. Realistic modeling of waste oxidation is needed to assure that the dose limits in 10 CFR 61.41 will be met. Adequate model support is essential to providing the technical basis for the model results.

2.3.4.3　　　Factor 2 - Hydraulic Isolation of Saltstone

Physical degradation of saltstone is expected to affect facility performance because more water can flow through a degraded waste form than an intact waste form, and increased water flow through the waste form is expected to increase radionuclide releases to ground water. Thus the physical degradation of saltstone during the 10,000-year performance period is of interest primarily because degradation is expected to compromise the hydraulic isolation of the waste.

Two important aspects of the NRC's plan to monitor the hydraulic isolation of saltstone are (i) to confirm that the hydraulic properties of saltstone at the disposal site are consistent with the properties of the laboratory samples of saltstone described in the waste determination and (ii) to monitor the development of better predictions of saltstone degradation over long time periods. Waste in one of the tanks, Tank 48, is unlike the rest of the salt waste at SRS because it contains a substantial amount of organic salts; as a result, staff expects to monitor the hydraulic properties and long–term degradation of saltstone made from this waste as a separate monitoring factor.

2.3.4.4 Factor 3 - Model Support

Adequate model support is essential to assessing whether the saltstone disposal facility can meet the requirements of 10 CFR 61.41. Essentially, model support provides assurance that the results of any models used to predict potential doses or intermediate results of submodels are consistent with independent data. In the TER, staff indicated it would monitor the development of model support in the following technical areas: (i) moisture flow through fractures in the concrete and saltstone located in the vadose zone, (ii) realistic modeling of waste oxidation and release of technetium, (iii) the extent and frequency of fractures in saltstone and vaults that will form over time, (iv) the plugging rate of the lower drainage layer of the engineered cap, and (v) the long-term performance of the engineering cap as an infiltration barrier. Implementation of an adequate erosion control design is important to ensuring that the provisions of 10 CFR 61.42 can be met. The erosion control barrier will help to maintain a thick layer of soil over the vaults, which reduces the potential for intrusion into the waste.

Each of these areas is related to other monitoring factors. However, the "model support" monitoring factor is different from the other factors because its goal is to provide confidence in aspects of the model or models used to make dose predictions. Thus, to monitor model support development, staff expects to compare available data about the development of the disposal system or analogous systems with model predictions. Ideally, model support includes multiple lines of evidence supporting the conclusions of modeled dose predictions or intermediate submodels, such as radionuclide release or transport in the subsurface. Lines of evidence may include site characterization and design data, results of process-level modeling, laboratory testing, field measurements, analogs, and formal independent peer review.

2.3.4.5 Factor 4 - Erosion Control Design

The infiltration and erosion controls are both part of an engineered cap that DOE plans to use to cover the saltstone disposal facility at facility closure. Implementation of an adequate erosion control design is important to protecting a potential inadvertent intruder, because the erosion control barrier will help to maintain a thick layer of soil over the vaults, which reduces the potential for intrusion into the waste. The staff plans to verify that the erosion control barrier is built as DOE described to the NRC during consultation or that, if changes are made to the design, the new design will be as effective in limiting erosion as the design described in documents used to support the waste determination.

2.3.4.6 Factor 5 - Infiltration Barrier Performance

The infiltration control system was identified as a factor for monitoring because the predicted dose to a potential member of the public was sensitive to DOE's assumption that the infiltration control system would significantly limit the amount of water reaching the waste for the entire 10,000–year performance period. To monitor the design and performance of the infiltration control system, staff expect to verify that the infiltration controls are implemented as described in the waste determination and supporting documents or that any changes made to the design do not degrade facility performance. Specifically, if the design is not changed, staff expects to monitor the development of information to support assumptions DOE made about the rate at which the lower drainage layer in the infiltration system would become plugged and any information developed to support the performance of the cap as an infiltration barrier.

2.3.4.7   Factor 6 - Feed Tank Sampling

Feed tank sampling is related to the final inventory of radionuclides in the saltstone disposal facility. Implementation of an adequate waste sampling plan is important to ensuring that the provisions of 10 CFR 61.41 and 10 CFR 61.42 can be met. It is necessary to confirm that the concentration of highly radioactive radionuclides (HRRs) in treated salt waste (or grout) is less than or equal to the concentration assumed in the waste determination. The staff expects to monitor how well each of the planned salt waste treatment processes removes radionuclides from the waste, because removal of radionuclides from the waste will affect the inventory of radionuclides in the salt waste disposal facility. In addition, staff will monitor radionuclide removal to assess whether potential doses to members of the general public will be maintained as low as reasonably achievable (ALARA), as required by the performance objective for protection of the general public from releases of radioactivity.

2.3.4.8   Factor 7 - Tank 48 Waste form

The chemical composition of the salt waste in Tank 48 differs from the salt waste in other tanks because it contains a substantial amount of organic salts. To ensure that Tank 48 waste can be safely managed, tests are needed to measure the physical properties of the waste form made from this waste to confirm that it will provide suitable performance. Staff plans to monitor reported disposal site inventories as well as sampling of the salt waste preparation feed tank to assess whether the inventory and concentrations of radionuclides sent to the saltstone disposal facility are consistent with the inventories and concentrations that DOE used as a basis for their waste determination.

2.3.4.9   Factor 8 - Removal Efficiencies

The removal efficiencies of HRRs by each of the planned salt waste treatment processes are a key factor in determining the radiological inventory disposed of in saltstone, which, in turn, is an important factor in determining that 10 CFR 61.41 and 10 CFR 61.42 can be met.

3.0 MONITORING AT THE IDAHO NATIONAL LABORATORY IDAHO NUCLEAR TECHNICAL AND ENGINEERING CENTER IN CALENDAR YEAR 2011

3.1 Introduction

In total, there are 15 waste storage tanks at the Tank Farm Facility (TFF) that include eleven 1,136 m^3 (300,000-gallon) tanks, four 114 m^3 (30,000-gallon) tanks, interconnecting transfer piping, and secondary containment components for the transfer piping. Placed into service between 1953 and 1966, the eleven 1,136 m^3 (300,000-gallon) tanks (WM-180 through WM-190) are approximately 15.2 m (50 ft) in diameter and 6.4-7.0 m (21-23 ft) in height. Nine of the eleven 1,136 m^3 (300,000-gallon) tanks are constructed of Type 304L stainless steel; two tanks (WM-180 and WM-181) use Type 347 stainless steel. Constructed in 1954, the four inactive 114 m^3 (30,000-gallon) stainless steel below-grade storage tanks, (WM-103 through WM-106), sit on reinforced concrete pads and were removed from service in 1983. The tanks are horizontal cylinders approximately 3.5 m (11.5 ft) in diameter and 11.6 m (38 ft) in length. All eleven 1,136 m^3 (300,000-gallon) tanks are housed in concrete vaults approximately 13.7 m (45 ft) below grade and the 114 m^3 (30,000-gallon) tanks do not have vaults.

The TFF has been used for the storage of a variety of radioactive wastes, including wastes directly from spent fuel reprocessing and other ancillary wastes since 1953. Spent fuel reprocessing wastes and other ancillary facility wastes were sent to the TFF until 1992.

Recent tank cleaning operations have resulted in the removal of the remaining sodium-bearing waste (SBW) and tank heels from seven 1,136 m^3 (300,000-gallon) tanks and four 114 m^3 (30,000-gallon) tanks. Four 1,136 m^3 (300,000-gallon) tanks remain to be cleaned, and these four tanks are anticipated to be cleaned as efficiently as the other 1,136 m^3 (300,000-gallon) tanks that have been cleaned. The residual waste inventories at closure in a stabilized form are expected to enable DOE to demonstrate that the TFF tank system residual waste at final closure will meet Section 3116 criteria. The TFF closure date is expected in 2012.

3.2 Background

On September 7, 2005, DOE submitted a draft waste determination for residual waste incidental to reprocessing, including sodium bearing waste, stored in the INTEC TFF to demonstrate compliance with the NDAA criteria including demonstration of compliance with the performance objectives in Part 61. In its consultation role, staff reviewed the draft waste determination and concluded that the NDAA criteria could be met for residual waste stored in the INTEC TFF. The NRC documented the results of its review in a technical evaluation report (TER) issued in October 2006 (NRC, 2006). DOE issued a final waste determination in November 2006 (DOE-Idaho, 2006) taking into consideration the assumptions, conclusions, and recommendations documented in NRC's TER.

To carry out its monitoring responsibilities under the NDAA, the NRC developed a monitoring plan for the INTEC TFF facility in April 2007 (NRC, 2007c) based on the risk-significant monitoring areas identified in the TER. The NRC conducted two onsite observations in 2007 to observe tank grouting operations (7 of 11 large tanks and 4 smaller tanks) at the INTEC TFF.

All open items identified in the first onsite observation conducted in April 2007 were closed in the August 2007 onsite observation.

In August 2008, staff participated in a third onsite observation to observe pipe grouting operations, radiation protection controls, and the environmental sampling program. No findings resulted from the three onsite observations. No tank farm closure activities occurred in CY 2009; therefore, staff elected to forego an onsite observation.

In CY 2010, staff made one site visit in August 2010 to conduct a tour of INL INTEC facilities (NRC, 2010a). During the visit, staff obtained updates on closure activities and schedules, met with state officials, and collected routine information related to several monitoring factors listed in the NRC's monitoring plan for the INTEC TFF, such as radiation protection and the environmental monitoring programs.

In CY 2011, the NRC did not make any observation visits to TFF because there were no active operations on site during the year. Although there were no onsite observations, staff did conduct technical reviews of two risk-significant areas identified in the TER, KMA 3 and KMA 4, as presented in Section 3.3.2. Appendix C provides a visual depiction of the timeline of NRC monitoring of the INTEC TFF facility under NDAA from 2007 to 2011.

3.3 NRC Monitoring Activities in 2011

3.3.1 Observation Visits

As mentioned above, in CY 2011, the NRC did not make any observation visits to TFF because there were no active operations on site during the year.

3.3.2 Technical Reviews

3.3.2.1 Technical Review Area for KMA 3

Key Monitoring Area 3 can be described as "hydrologic uncertainty":

> "Relevant recent and future monitoring data and modeling activities should continue to be evaluated to ensure that hydrological uncertainties that may significantly alter the conclusions in the PA and TER are addressed. If significant new information is found, this information should be evaluated against the PA and TER conclusions..." (Description of KMA 3; see Table B-2)

KMA 3 was developed as a result of staff's review of the INTEC TFF draft waste determination and supporting PA as documented in NRC (2006), which showed a number of uncertainties associated with DOE's ground water model used to support its demonstration of compliance with the performance objective found in 10 CFR 61.41 for protection of the general population from releases of radioactivity. Some of the largest hydrogeological uncertainties impacting facility performance were related to infiltration rates and the impact of Big Lost River seepage on contaminant releases from the tank farm. Nonetheless, staff was able to conclude with reasonable assurance that natural system uncertainty could be managed with conservative assumptions. In other words, given the large safety margin between the performance standard of 0.25 mSv/yr (25 mrem/yr) and DOE's estimated peak dose of 0.005 mSv/yr) 0.5 mrem/yr for

3-2

the INTEC TFF, less natural system performance was needed than was taken by DOE in its PA to demonstrate compliance. For example, more easily supportable dilution factors attributable to mixing in the Snake River Plain Aquifer alone for key radionuclides such as technetium-99 and iodine-129 was found to be sufficient for DOE to demonstrate compliance with 10 CFR 61.41.

As stated in the monitoring plan for the INTEC TFF (NRC, 2007c), staff planned to continue to stay abreast of relevant monitoring and modeling activities conducted by DOE, other agencies, or independent researchers until such time that NRC staff could confidently conclude that overall system performance was adequately studied and constrained. If issues related to engineered barrier system performance arose during evaluation of KMA 2, then KMA 3 would become increasingly important. Therefore, NRC staff determined that the status of this KMA would remain open until KMA 2 was closed.

NRC staff typically reviews ground water-monitoring reports related to the INTEC facility conducted under the CERCLA program. Data from historical releases collected under the CERCLA program is helpful to staff with respect to evaluating hydrogeological system uncertainties. It is important to note that risks associated with historical releases are addressed under the CERCLA program and are not considered when evaluating potential compliance with performance objectives under the NDAA (i.e., only future releases associated with or following tank closure are considered when evaluating compliance with 10 CFR Part 61 performance objectives). Thus, CERCLA information is reviewed for the sole purpose of providing risk insights on future natural system performance rather than as a measure of contemporaneous compliance with performance objectives for LLW disposal under the NDAA.

DOE Idaho prepares an annual report (e.g., DOE- Idaho, 2011b) describing maintenance, inspection, and other activities performed to address contaminated soils and ground water at INTEC as specified in the Record of Decision for the Tank Farm Soil and INTEC Ground water Operable Unit 3-14, signed in May 2007 (DOE- Idaho, 2007). However, DOE's annual reports are not intended to interpret data, form conclusions, or determine the effectiveness of the selected remedy; these topics are the subject of DOE's 5-year review of the effectiveness of its CERCLA response actions. A 5-year review was recently completed and documented in a January 2011 report (DOE- Idaho, 2011a). This report is discussed further below.

Current risks associated with tank farm soil and INTEC ground water from previous releases include external exposure to soil contaminated with cesium-137 and ingestion of contaminated Snake River Plain aquifer (SRPA) ground water. The SRPA currently contains significant concentrations of strontium-90 and nitrate from previous injection well operations and technetium-99 resulting from tank farm releases (DOE-Idaho, 2011b). If left unmitigated, perched water could become a continuing source of ground water contamination to the SRPA above certain CERCLA action levels (e.g., maximum contaminant levels or MCLs) beyond 2095. CERCLA modeling shows that with decreased infiltration in a 3.8-hectare (9.5-acre) area surrounding the Tank Farm Facility, the SRPA could meet action levels by 2095. This 3.8-hectare (9.5-acre) area is designated a recharge control zone under the selected remedy. Thus, remedial activities are focused on the control of recharge to the subsurface.

DOE's 2010 annual monitoring report (DOE Idaho, 2011b) describes various activities designed to control infiltration including inspection activities, remedial actions (e.g., laying down asphalt over decommissioned areas; constructing and lining ditches), identification of anthropogenic sources of water, plugging abandoned wells, etc. Section 5 of DOE's annual monitoring report describes long-term monitoring activities that are of particular interest to NRC staff in its review of KMA 3. During the FY 2010 reporting period DOE conducted ground water sampling at 14 SRPA wells[5] and five additional wells sampled as part of the Idaho CERCLA Disposal Facility monitoring program[6]. The operating unit monitoring plan requires sampling of 15 aquifer wells during the even years. One well, USGS-57, was not sampled in FY2010 because of an inoperable submersible pump at the time of the sampling event. Perched water samples were collected from six perched wells: 55-06, ICPP-2018, ICPP-2019, MW-2, MW-5-2, MW-10-2. Well 33-1 was not sampled because the well contained insufficient water for sampling. Well MW-10-2 only had enough water for a partial suite of analyses. Figure 3-1 shows locations of monitoring wells.

Consistent with previous data, the highest technetium-99 concentrations from the April 2010 sampling event were associated with monitoring well ICPP-MON-A-230 (71 kBq/m^3 or 1,930 pCi/L) located near the INTEC Tank Farm and the second-highest technetium-99 concentrations were measured at aquifer well ICPP-2021 (50 kBq/m^3 or 1,340 pCi/L), located southeast of the Tank Farm (see Figure 3-1). These two wells were the only wells to exceed the technetium-99 MCL[7] of 33 kBq/m^3 (900 pCi/L). All wells show stable or declining trends.

Consistent with previous data, very high strontium-90 levels (>370 kBq/m^3 or 10,000 pCi/L) were observed in the northern shallow perched water across INTEC. The highest strontium-90 concentrations were observed in wells southeast of the Tank Farm. The maximum strontium-90 concentration detected was 5.7 MBq/m^3 (154,000 pCi/L) at monitoring well ICPP-2018 (see Figure 3-1). At most well locations, strontium-90 concentrations were similar to those observed during the previous year, but are approximately half those reported in the same wells during the mid-1990s because of decay and transport. Gross beta activity was detected at nearly all perched water sampling locations with the highest gross beta level occurring at well ICPP-2018 (12 MBq/m^3 or 326,000 pCi/L) consistent with the strontium-90 data. Strontium-90 was detected in 13 of 14 SRPA wells with samples from seven of the wells exceeding the Sr-90 MCL of 296 Bq/m^3 (8 pCi/L). The highest measurement of Sr-90 in the aquifer was 918 Bq/m^3 (24.8 pCi/L) at well USGS-47 located down gradient of the former INTEC injection well. All wells showed similar or slightly lower strontium-90 levels compared to the previous reporting period.

Although uranium-234 and uranium-238 were present at background levels, no detectable gross alpha levels were reported for any perched water sampling locations. This apparent discrepancy may be explained by the lower detection limit for uranium isotopes compared to gross alpha of 18.5 and 148 Bq/m^3 (0.5 and 4 pCi/L), respectively. Analysis revealed no detectable levels of plutonium in the vadose zone and aquifer wells.

The lateral extent of the northern shallow perched water was also mapped in the FY 2010 report. Changes in water levels at several wells could be attributable to contributions from or elimination of anthropogenic sources of water. The Big Lost River (BLR) is another potential source that can impact perched water levels at INTEC and flowed past INTEC between June 9 and 14 and between June 17 and 20, 2010. Similar to previous observations, only one

---

[5]SRPA wells CPP-01, ICPP-2020, ICPP-2021, ICPP-MON-A-1230, MW-18-4, USGS-040, USGS-41, USGS-42, USGS-47, USGS-48, USGS-51, USGS-52, USGS-59, USGS-067 were sampled in the April 2010 event.

[6] Wells ICPP-1782, ICPP-1783, ICPP-1800, ICPP-1829, and ICPP-1831 were sampled in the April 2010 event.

[7] Note that the NRC does not use MCLs or maximum contaminant levels to determine compliance with performance objectives in 10 CFR Part 61, Subpart C. MCLs are standards used by the U.S. Environmental Protection Agency in the CERCLA program and are provided for information only.

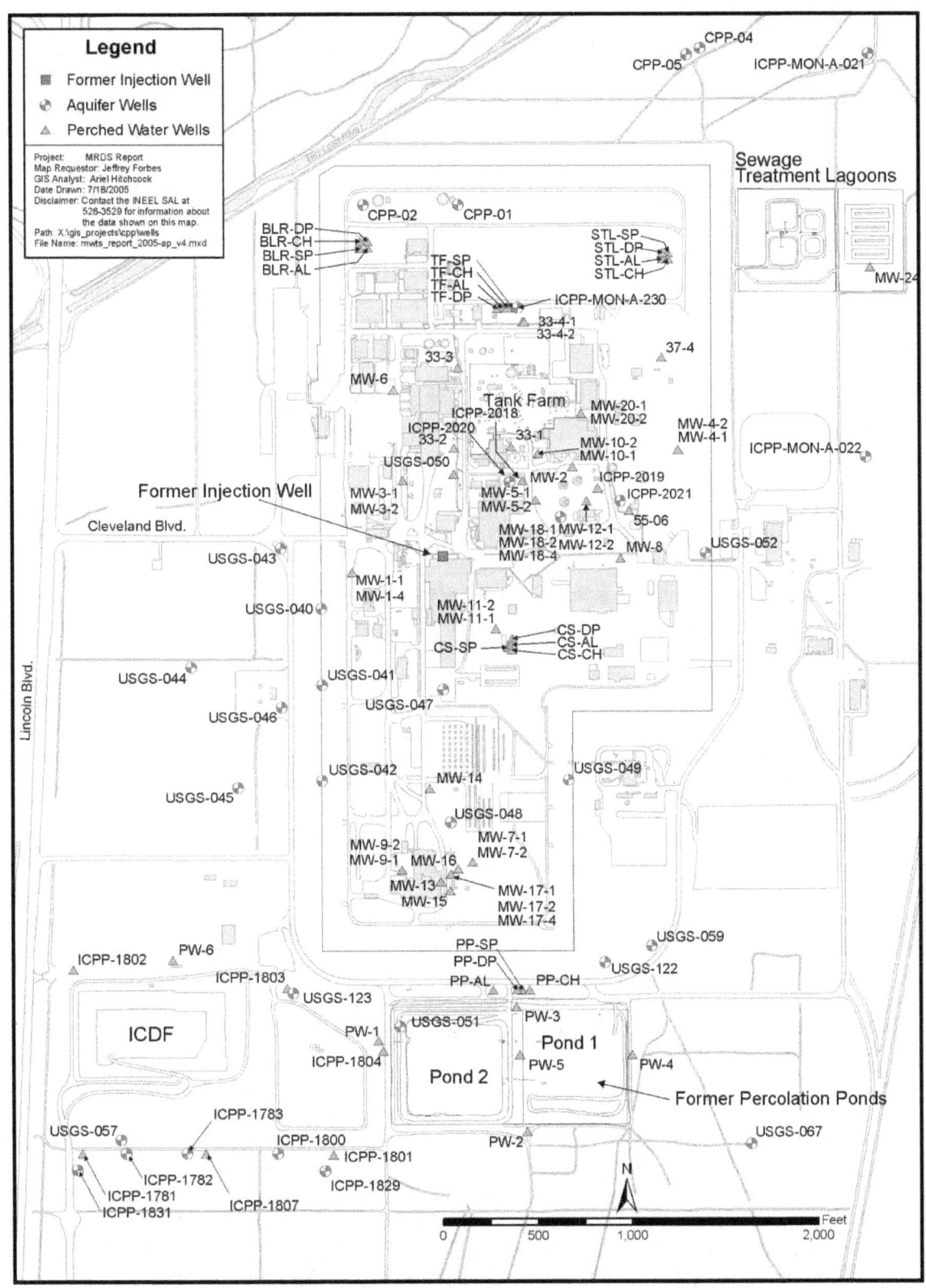

Figure 3-1: INTEC TFF monitoring-well network (from DOE-Idaho, 2011b)

monitoring well (Well BLR-CH) showed a significant water-level response to the river flow event. Well BLR-CH is the well closest to the river (i.e., 152 m [500 ft] from the river channel). After a 4-day time lag following the onset of flow in the river, the perched water level in Well BLR-CH rose 3 m (10 ft) over 15 days. This is similar to the water-level response observed in the past at this well location. No other wells showed any response to flow changes in the river.

DOE also conducts a 5-year review of CERCLA response actions. This 5-year review was recently conducted for the INTEC and documented in a report issued in January 2011 (DOE Idaho, 2011a). With respect to perched water and ground water, DOE concluded that the CERCLA response actions were functioning as intended and that previous exposure assessment assumptions remain valid. Since the 2007 ROD, DOE indicates that significant progress has been made towards reducing precipitation infiltration and anthropogenic recharge at INTEC. Plans to install a low permeability cover over the tank farm and surrounding area will proceed as facilities are decommissioned. Although remedial activities are not yet complete and their ultimate effectiveness cannot be assessed at this time, DOE concludes that indications are favorable that the desired effect of these remedies will be achieved. Staff agrees with this assessment.

During FY 2010, DOE contractors also performed a modeling analysis that addressed an NRC staff recommendation made during NRC's 2010 onsite observation (Recommendation 2010-2). Staff recommended that DOE consider (in its decision to update the PA during execution of its periodic PA maintenance review) recent data collected under the CERCLA program that appears to be inconsistent with the DOE PA modeling results with respect to the impact of BLR flow on contaminant fate and transport at the INTEC TFF. Recent observations of limited perched water level response in vadose zone wells following BLR flow and other investigations indicate that anthropogenic sources of water associated with INTEC operations, rather than BLR seepage, are a more significant source of perched water currently observed at INTEC TFF. Ultimately, DOE determined that this issue was significant enough to include in its PA maintenance checklist and performed additional modeling to determine the potential dose impact for more vertical movement of water in the vadose zone at the INTEC TFF, in comparison to DOE's PA model that indicates significant lateral spread and dilution of contaminants that might be released from the tanks to the INTEC vadose zone in the future.

Staff reviewed DOE's modeling analysis (Portage, 2011) that showed while the doses would increase by roughly a factor of two, performance objectives could still be met. DOE's analysis was conducted using the DUST-MS code originally used to develop a source term in the INTEC TFF PA. The source term was used as input to the GWSCREEN simulations that were used to simulate vadose and saturated zone transport. Because flow through the TFF vadose zone was assumed to be vertical in this alternative conceptual model (along with 1-D flow in the saturated zone), a multi-dimensional model such as PORFLOW was not needed to perform the ground water simulations. In general, the supplemental analysis appears to be technically sound. However, it is interesting to note that DOE's PORFLOW simulations used to prepare the INTEC TFF PA indicated that vadose zone dilution would be rather significant (i.e., concentrations released from the tanks would be thousands to tens of thousands times less during transport through the vadose zone). Presumably, dilution in the vadose zone in the PA modeling was almost entirely offset by dilution and dispersion in the SRPA during transport to the 100 m well location in the supplemental analysis. It would be helpful for DOE to further explain the performance impact associated with the alternative conceptual model and any offsets between vadose and saturated zone dilution to further support the revised estimates. Key modeling parameters such as Darcy velocity, effective porosity, dispersivity, etc. should be fully supported and a sensitivity analysis conducted to study the impact of parameter uncertainty on dose predictions.

Staff identified no new and significant information that would invalidate it's TER conclusions. Information on infiltration rates and the mobility of radiological constituents will continue to be assessed by staff through review of INTEC monitoring data and other sources of information. BLR seepage near the INTEC TFF will also continue to be evaluated to determine its potential impact on ground water flow and transport mechanisms near the TFF. Staff continues to have reasonable assurance that performance objectives will be met for the INTEC TFF facility.

Staff also continues to recommend the following:

Recommendation 2007-3: NRC staff recommends that DOE evaluate any new and significant information related to hydrogeological system uncertainty at INTEC. NRC requests that DOE provide any recent reports or data related to hydrogeological system uncertainty at INTEC to NRC for review as that information becomes available.

3.3.2.2   Technical Review Area for KMA 4

Key Monitoring Area 4 can be described as "monitoring during operations":

"Closure and post-closure operations (until the end of active institutional controls, 100 years) will be monitored to ensure that the §61.43 performance objective (protection of individuals during operations) can be met. As part of this assessment radiation records, environmental monitoring, and exposure assessment calculations may be reviewed." [Description of KMA 4; see Table B-2]

KMA 4 in the NRC's TER for INTEC TFF addresses DOE compliance with the performance objective found in 10 CFR 61.43 related to protection of individuals during operations[8]. To evaluate this performance objective, the INL monitoring plan provides that staff will review DOE worker radiation records, DOE's program to maintain worker doses ALARA, and offsite dose assessment methods and results. Technical review activities associated with protection of members of the public under KMA 4 discussed in this section include the review of environmental surveillance data and analyses performed by Gonzales Stoller Surveillance, LLC (formerly S.M. Stoller Corporation) and Idaho Department of Environmental Quality (Idaho DEQ).

Current activities at the INTEC TFF include storage of spent nuclear fuel (SNF) in a modern water basin and in dry storage facilities, management of high-level waste calcine and sodium-bearing liquid waste, and the operation of the Idaho Comprehensive Environmental Response, Compensation, and Liability Act (CERCLA) Disposal Facility (ICDF), which includes a landfill, evaporation ponds, and a storage and treatment facility. Although various activities, including the demolition of 31 structures previously associated with the grouted tanks occurred at the site, no major closure activities that may impact the dose to workers and members of the public occurred at the INTEC TFF during CY 2010.

Staff collected and reviewed monitoring data from DOE's 2010 environmental surveillance reports, the Idaho DEQ INL Oversight Program annual report for calendar year 2010, and Idaho

---

[8] Effluents from operational activities are limited to 0.25 mSv (25 mrem) for doses to the public, in accordance with 10 CFR 61.41.

DEQ's quarterly surveillance reports for the first and second quarters of 2011. Staff used this information to evaluate the impacts of INL operations on members of the public as well as evaluate the air, soil, water, vegetation, animals, and foodstuffs on and around the INL site to confirm compliance with applicable laws and regulations. Since these reports cover the entire site and are not focused specifically on the INTEC TFF (which is a subset of the entire site), the NRC considers these analyses to be bounding dose analyses for releases to the public.

The DOE-Idaho environmental surveillance program, which performs monitoring activities on the INL Site, at the INL Site boundary, and offsite emphasizes the measurement of airborne radionuclides because the air transport pathway is considered to be the principal pathway from the INL site for potential releases to the public. Results show that all radionuclide concentrations in ambient air samples were below DOE standards and within historical measurements and are considered to have no measurable impact on the environment. Two different computer programs were used to estimate doses. The Clean Air Act Assessment Package, 1988 (CAP-88), computer code was used to calculate the dose to the hypothetical, maximally exposed individual (MEI) and the mesoscale diffusion (MDIFF) air dispersion model was used to estimate the dose to the population within 80 km (50 miles) of the INL site facilities. The maximum dose to the MEI was calculated to be 5.8×10^{-4} mSv/year (0.058 mrem/year), well below the applicable radiation protection standard of 0.1 mSv/year (10 mrem/year). For comparison, the dose from natural background radiation was estimated to be 3.82 mSv (382 mrem). The maximum potential population dose to the approximately 306,000 people residing within a 80 km (50 mile) radius of any INL Site facility was calculated as 1.62×10^{-2} person-Sv (1.62 person-rem), below that expected from exposure to background radiation (1,168 person-Sv or 116,868 person-rem).

Surface water and ground water pathways are not considered to be major contributors to the public dose. Radionuclide measurements associated with surface water and ground water sources continue to be consistent with historical data, remaining well below the 0.04 mSv/yr (4 mrem/yr) EPA standard for public drinking water systems.

The maximum potential individual doses from consumption of waterfowl and big game animals from the INL site were estimated from the highest concentrations of radionuclides measured in samples collected at the site. Current trends show that these doses are lower than the maximum dose estimates from previous periods. The maximum potential dose of 6×10^{-4} mSv (6×10^{-2} mrem) for waterfowl samples is well below the 8.9×10^{-3} mSv (0.89 mrem) estimated from the most contaminated ducks, collected between 1993 and 1998 from sewage lagoons adjacent to the radioactive wastewater ponds. It is assumed that the ducks used the radioactive wastewater lagoons while in the area. The potential dose from consumption of meat from big game animals was estimated to be approximately 4×10^{-5} mSv (4×10^{-3} mrem). Although considered in the past, contributions from the game animal consumption pathway to population dose are not considered because only a limited percentage of the population hunts game, few of the animals killed have spent time on the INL Site, and most of the animals that do migrate from the INL site have low concentrations of radionuclides in their tissues by the time they are harvested. In general the dose contributions from the game animal consumption pathway can be expected to be less than the sum of the population doses from inhalation of air, submersion in air, ingestion of vegetables, and deposition on soil. Based on the graded approach used to evaluate nonhuman biota it can also be concluded that there is no evidence that INL site-related radioactivity associated with the soil or water is harming the resident plant and animal populations.

Staff also reviewed environmental data collected by the State of Idaho. The Idaho DEQ maintains an environmental surveillance program that analyzes samples (e.g., air, water [surface and ground water], soil, and milk) on and around the outside of the INL Site to help independently evaluate DOE's monitoring program and to assess potential environmental impacts from INL facilities. Idaho DEQ collects, analyzes, and publishes monitoring data in quarterly reports as well as an annual report. These reports are posted on Idaho DEQ's INL Oversight website (see http://www.deq.idaho.gov/inl_oversight). Staff has concluded that Idaho DEQ's independent environmental surveillance program is sufficient to support its annual review and plans to continuously review data, analyses, and conclusions provided in Idaho DEQ quarterly and annual reports to help reach its conclusions regarding compliance with the 10 CFR 61.43 performance objective.

Staff reviewed the 2010 annual report as well as the quarterly reports for calendar year 2010 and the first and second quarters of 2011 to determine potential offsite impacts to members of the public, unexplained or unexpected releases of radioactivity because of operations at INTEC, as well as to identify trends with respect to contaminant concentrations from onsite monitoring wells. While the monitoring network at INTEC is not as extensive as it is for the CERCLA program, onsite ground water monitoring data collected by Idaho DEQ does help to validate data collected by DOE and its contractors. Data reported in the 2010 annual report (Idaho DEQ, 2011c) and the 2011 quarterly reports for the first (Idaho DEQ, 2011b) and second quarters (Idaho DEQ, 2011a) were generally consistent with historic trends. Concentrations of radioactivity in air, soil, and milk samples were consistent with background levels. Radiation levels were also consistent with historic background measurements. In general, there appears to be good agreement between the environmental monitoring data reported by Idaho DEQ and data collected by DOE and its contractors.

Staff thinks that the consistency between data collected by Idaho DEQ and DOE provides confidence that both programs can be used to evaluate offsite environmental impacts associated with INL operations. Based, in part, on the environmental surveillance data collected by DOE and the State, NRC staff continues to have reasonable assurance that the 10 CFR 61.43 performance objective related to protection of individuals during operations will be met.

Staff will continue to evaluate worker and public exposure data or estimates through review of worker radiation records and review of environmental surveillance reports as the INTEC TFF closure activities progress in support of the technical review activities identified for KMA 4 in the INL monitoring plan (NRC, 2007c). The level of monitoring is expected to be higher during active closure operations conducted through the year 2012.

Recommendation 2007-4: DOE should provide information on any violations of requirements related to workers and the general public (10 CFR Part 835 or DOE Order 5400.5) during its waste disposal operations. As information provided on the Web may not be timely, staff requests that DOE provide information regarding worker or public dose exceedances within a reasonable timeframe of their occurrence.

3.3.3   Summary of Open Issues, Follow-up Actions, and Recommendations

Based on the August 10, 2010, observation trip, staff made two recommendations for DOE to consider in its decision to update the PA (NRC, 2010a).

Staff recommended:

Recommendation 2010-1: NRC staff recommended that the PA reflect the results of simulations performed and additional documentation generated during the NDAA consultation process to answer NRC staff inquiry regarding the cause and performance impact of the significant lateral spread of the contaminant plume emanating from the TFF to the south (e.g., caused by pressure gradient from BLR and resulted in up to a factor of 10,000 decrease in contaminant concentrations emanating from the TFF for relatively mobile [non-sorbing] constituents such as technetium-99 and I-129).

Recommendation 2010-2: NRC staff also recommended that DOE consider (in its decision to update the PA) recent data collected under the CERCLA program that appears to be inconsistent with the DOE PA modeling results with respect to the impact of BLR flow on contaminant fate and transport at the INTEC TFF.

As discussed in Section 3.3.2.1, DOE conducted an analysis in response to Recommendation 2010-2. Staff reviewed DOE's modeling analysis (Portage, 2011), which showed that while the doses would increase by roughly a factor of two, performance objectives could still be met. Staff notes that it would be helpful for DOE to further explain the performance impact associated with the alternative conceptual model and any offsets between vadose and saturated zone dilution to further support the revised estimates. Key modeling parameters such as Darcy velocity, effective porosity, dispersivity, etc. should be fully supported and a sensitivity analysis could be conducted to study the impact of parameter uncertainty on dose predictions.

There are no new open issues or recommendations for INL from CY 2011. Based on the analysis conducted by DOE and the NRC's review of documentation, NRC staff is confident that the current radiation protection program at INTEC TFF can meet the performance objectives as stated in 10 CFR 61.43 and 61.41. DOE provided proper documentation to demonstrate that activities were being conducted in a manner that is protective of individuals during operations.

4.0   SUMMARY OF ALL OPEN ISSUES AND RECOMMENDATIONS FOR SALTSTONE-SRS AND INL-TFF

Table 4-1 and Table 4-2 summarize the open issues and recommendations, respectively, which staff identified during its ongoing monitoring of DOE waste disposal actions from January 1, 2007, through December 31, 2011, under NDAA.

An issue is opened during monitoring activities for items identified by staff of higher risk-significance than follow-up actions. Open issues require additional follow-up by the NRC staff or additional information from DOE to address questions that the NRC staff has raised regarding DOE disposal actions.

A recommendation is an NRC suggestion to DOE to address potential issues identified during monitoring and usually results from a follow-up action. By their nature, recommendations do not require follow-up and they are not considered open or closed.

Table 4-1: Summary Description of Open Issues for CY 2011 in the NRC Section 3116(b) Monitoring Program

Open Issues		
Number	**Description**	**Status**
2007-1	At the SRS Saltstone facility, as a result of variations in the composition of saltstone grout actually produced at the SRS SPF, DOE should determine the hydraulic and chemical properties of as-emplaced saltstone grout. Inadequate saltstone grout quality could result in disposal actions that are not compliant with the 10 CFR 61.41 performance objective.	Open
2007-2	At the SRS Saltstone facility, DOE should demonstrate that intrabatch variability, flush water additions to freshly poured saltstone grout at the end of each production run, and additives used to ensure processability are not adversely affecting the hydraulic and chemical properties of the final saltstone grout. DOE should show that hydraulic and chemical properties are consistent with the assumptions in the waste determination or show that any deviations are not significant with respect to demonstrating compliance with the performance objectives.	Open
2009-1	At the SRS Saltstone facility, DOE should demonstrate that (1) technetium-99 in salt waste is converted to its reduced chemical form in saltstone grout during the curing of saltstone grout and is thereby strongly retained in saltstone grout, and (2) the sorption of dissolved technetium-99 onto saltstone grout and vault concrete is consistent with the K_d values for technetium-99 assumed in the performance assessment.	Open

Table 4-2: Summary of Staff Recommendations for CY 2011 under the NRC Section 3116(b) Monitoring Program

Recommendations	
Number	**Description**
2007-3	At the INL INTEC TFF, NRC staff recommends that DOE evaluate any new and significant information related to hydrogeological system uncertainty at INTEC. NRC requests that DOE provide any recent reports or data related to hydrogeological system uncertainty at INTEC to NRC for review as that information becomes available.
2007-4	At the INL INTEC TFF, DOE should provide information on any violations of requirements related to workers and the general public (10 CFR Part 835 or DOE Order 5400.5) during its waste disposal operations. As information provided on the Web may not be timely, NRC staff requests that DOE provide information regarding worker or public dose exceedances within a reasonable timeframe of their occurrence.

Table 4-3: Summary of Follow-Up Actions for CY 2011 under the NRC Section 3116(b) Monitoring Program

Follow-Up Actions	
Number	**Description**
2011-1	At the SRS Saltstone facility, Disposal Unit 2 Water Tightness Test Quality Assurance Records: DOE will provide NRC staff with documentation of cell design changes and hydrotesting results for review when they are available following the Operational Readiness Review.
2011-2	At the SRS Saltstone facility, Anchor Bolt Penetrations in Vault 4/Vault 4 Floor: Seepage has occurred at imperfections in the vault walls as liquid builds up in the gap between the saltstone and vault wall. DOE has applied sealant coatings, a rain shield, certified huts, and a drip pan on the exterior of the vault cells to reduce seepage of liquid to the environment. This follow-up action remains open pending the response to the above concern and the completion of the work DOE is performing on this follow-up action. The NRC staff will continue to review documentation regarding this follow-up action as it becomes available.

5.0   REFERENCES

10 CFR Part 61. Title 10 of the *Code of Federal Regulations*, Part 61, "Licensing Requirements for Land Disposal of Radioactive Waste," 1982.

10 CFR Part 20. Title 10 of the *Code of Federal Regulations,* Part 20, "Standards for Protection against Radiation," 1991.

10 CFR Part 835. Title 10 of the Code of Federal Regulations, *Title 10,* Part 835, "Occupational Radiation Protection," 1993.

CNWRA and NRC, 2011. Center for Nuclear Waste Regulatory Analyses and Nuclear Regulatory Commission, "Experimental Study of Contaminant Release from Reducing Grout", R.T. Pabalan, G.W. Alexander, and D.J. Waiting, March 29, 2012, ADAMS Accession No. ML12089A319.

CNWRA , SWRI, and NRC, 2011. Center for Nuclear Waste Regulatory Analyses, Southwest Research Institute, and Nuclear Regulatory Commission, "Bonding And Cracking Behavior And Related Properties Of Cementitious Grout In An Intermediate-Scale Grout Monolith", September 2011, ADAMS Accession No. ML112700061.

DOE, 2005. U.S. Department of Energy, "Draft Basis for Section 3116 Determination Salt Waste Disposal at the Savannah River Site," DOE-WD-2005-001, Washington, DC, March 2005.

DOE, 2006. U.S. Department of Energy, "Basis for Section 3116 Determination Salt Waste Disposal at the Savannah River Site," DOE-WD-2005-001, Washington, DC, January 2006.

DOE, 2009. U.S. Department of Energy, "*Performance Assessment for the Saltstone Disposal Facility at the Savannah River Site*", SRR-CWDA-2009-00017, Savannah River Site, Aiken, SC, October 29, 2009, ADAMS Accession No. ML101590008.

DOE, 2010. U.S. Department of Energy, "Saltstone Inadvertent Transfer on May 19, 2010 Presentation", SRR-LWO-2010-00047, Savannah River Site, Aiken, SC, July 2010, ADAMS Accession No. ML102180304.

DOE-Idaho, 2006. U.S. Department of Energy Idaho Operations Office, "Basis for Section 3116 Determination for the Idaho Nuclear Technology and Engineering Center Tank Farm Facility," DOE/ID–11226, Rev. 0, Idaho Falls, ID, November 2006.

DOE-Idaho, 2007. U.S. Department of Energy Idaho Operations Office, "Record of Decision for Tank Farm Soil and Idaho Nuclear Technology and Engineering Center Ground water, Operable Unit 3-14," DOE/ID–11296, Rev. 0, Idaho Falls, ID, May 16, 2007.

DOE-Idaho, 2011a. U.S. Department of Energy Idaho Operations Office, "Five-Year Review of CERCLA Response Actions at the Idaho National Laboratory Site—Fiscal Years 2005-2009," DOE/ID-11429, Rev. 0, Idaho Falls, ID, January 2011.

DOE-Idaho, 2011b. U.S. Department of Energy Idaho Operations Office, "Fiscal Year 2010 Annual Operations and Maintenance Report for Operable Unit 3-14, Tank Farm Soil and INTEC Ground water," DOE/ID-11442, Rev. 0, Idaho Falls, ID, August 2011.

DOE-Idaho, 2011c. U.S. Department of Energy Idaho Operations Office, "Idaho National Laboratory Site Environmental Report Calendar Year 2010", DOE/ID-12082(10) ISSN 1089-5469 GSS-ESER-151, Environmental Surveillance, Education and Research Program, September 2011. http://www.gsseser.com/Annuals/2010/PDFs/2010ASERReport.pdf

Idaho, DEQ, 2011a. "Environmental Surveillance Program Quarterly Report, April – June 2011," Department of Environmental Quality, Idaho National Laboratory Oversight Program. http://www.deq.idaho.gov/inl_oversight/library/2011_env_surv_q2.pdf

Idaho DEQ, 2011b. "Environmental Surveillance Program Quarterly Report, January – March, 2011," Department of Environmental Quality, Idaho National Laboratory Oversight Program. http://www.deq.idaho.gov/inl_oversight/library/2011_env_surv_q1.pdf

Idaho DEQ, 2011c. "DEQ-INL Oversight Program Annual Report 2010," Department of Environmental Quality, Idaho National Oversight Program. http://www.deq.idaho.gov/media/781813-inl-oversight-annual-report-2010.pdf

ICRP, 1977. International Commission on Radiological Protection, "Recommendations of the International Commission on Radiological Protection," Publication 26, Annals of the ICRP 1 (3) 1977.

NRC, 2005a. U.S. Nuclear Regulatory Commission, "Technical Evaluation Report for Draft Waste Determination for Salt Waste Disposal," Letter from L. Camper to C. Anderson, DOE, December 28, 2005, ADAMS Accession No. ML053010225.

NRC, 2005b. U.S. Nuclear Regulatory Commission, Staff Requirements Memorandum for SECY-05-0073, "Implementation of New NRC Responsibilities under the National Defense Authorization Act of 2005 in Reviewing Waste Determinations for the USDOE," Washington, DC, U.S. Government Printing Office, June 30, 2005, ADAMS Accession No. ML051810375.

NRC, 2006. U.S. Nuclear Regulatory Commission, "*Nuclear Regulatory Commission Technical Evaluation Report for the U.S. Department of Energy Idaho National Laboratory Site Draft Section 3116 Waste Determination for Idaho Nuclear Technology and Engineering Center Tank Farm Facility,*" Washington, DC, October 2006, ADAMS Accession No. ML062490142.

NRC, 2007a. U.S. Nuclear Regulatory Commission, "NRC Staff Guidance for Activities Related to U.S. Department of Energy Waste Determinations: Draft Final Report for Interim Use," NUREG-1854, Washington, DC, August 2007, ADAMS Accession No. ML072360184.

NRC, 2007b. U.S. Nuclear Regulatory Commission, "Nuclear Regulatory Commission Plan for Monitoring the U.S. Department of Energy Salt Waste Disposal at the Savannah River Site in Accordance with the National Defense Authorization Act for Fiscal Year 2005," Washington, DC, May 3, 2007, ADAMS Accession No. ML070730363.

NRC, 2007c. U.S. Nuclear Regulatory Commission, "Nuclear Regulatory Commission Plan for Monitoring the Disposal Actions Taken by the U.S. Department of Energy at the Idaho National Laboratory Idaho Nuclear Technology and Engineering Center Tank Farm Facility in Accordance with the National Defense Authorization Act for Fiscal Year 2005," Washington, DC, April 13, 2007, ADAMS Accession No. ML070650222.

NRC, 2008a. U.S. Nuclear Regulatory Commission, "Nuclear Regulatory Commission October 29-30, 2007 Onsite Observation Report for the Savannah River Site Saltstone Facility," January 31, 2008, ADAMS Accession No. ML073461038.

NRC, 2008b. U.S. Nuclear Regulatory Commission, , "NRC Periodic Compliance Monitoring Report for U.S. Department of Energy Non-High-Level Waste Disposal Actions, Annual Report for Calendar Year 2007," NUREG-1911, Washington, DC, August 2008, ADAMS Accession No. ML082280145.

NRC, 2008c. U.S. Nuclear Regulatory Commission, "Nuclear Regulatory Commission March 24-28, 2008 Onsite Observation Report for the Savannah River Site Saltstone Facility," June 5, 2008, ADAMS Accession No. ML081290367.

NRC, 2009a. U.S. Nuclear Regulatory Commission, "Nuclear Regulatory Commission March 25-26, 2009 Onsite Observation Report for the Savannah River Site Saltstone Facility," May 22, 2009, ADAMS Accession No. ML091320439.

NRC, 2009b. U.S. Nuclear Regulatory Commission, "NRC Periodic Compliance Monitoring Report for U.S. Department of Energy Non-High-Level Waste Disposal Actions, Annual Report for Calendar Year 2008," NUREG-1911, Rev 1, Washington, DC, May 2009, ADAMS Accession No. ML091400501.

NRC, 2010a. U.S. Nuclear Regulatory Commission, "Nuclear Regulatory Commission August 10, 2010 Onsite Observation Report for the Idaho National Laboratory Idaho Nuclear Technology and Engineering Center Tank Farm Facility," October 14, 2010, ADAMS Accession No. ML102770022.

NRC, 2010b. U.S. Nuclear Regulatory Commission, "Nuclear Regulatory Commission July 28, 2010 Onsite Observation Report for the Savannah River Site Saltstone Facility," November 19, 2010, ADAMS Accession No. ML102180250.

NRC, 2010c. U.S. Nuclear Regulatory Commission, "Nuclear Regulatory Commission Second Request for Additional Information on the 2009 Performance Assessment for the Saltstone Disposal Facility at the Savannah River Site," December 15, 2010, ADAMS Accession Number ML103400571.

NRC, 2010d. U.S. Nuclear Regulatory Commission, "NRC Periodic Compliance Monitoring Report for U.S. Department of Energy Non-High-Level Waste Disposal Actions, Annual Report for Calendar Year 2009," NUREG-1911, Rev 2, Washington, DC, August 2010, ADAMS Accession No ML102571453.

NRC, 2011a. U.S. Nuclear Regulatory Commission, "Nuclear Regulatory Commission January 27, 2011 Onsite Observation Report for the Savannah River Site Saltstone Facility," March 15, 2011, ADAMS Accession No. ML110590941.

NRC, 2011b. U.S. Nuclear Regulatory Commission, "Nuclear Regulatory Commission April 26, 2011 Onsite Observation Report for the Savannah River Site Saltstone Facility," August 19, 2011, ADAMS Accession No. ML111890319.

NRC, 2012a. U.S. Nuclear Regulatory Commission, "Technical Evaluation Report for the Revised Performance Assessment for the Saltstone Disposal Facility at the Savannah River Site, South Carolina," April 30, 2012, ADAMS Accession Number ML121170309.

NRC, 2012b. U.S. Nuclear Regulatory Commission, "NRC Periodic Compliance Monitoring Report for U.S. Department of Energy Non-High-Level Waste Disposal Actions, Annual Report for Calendar Year 2010," NUREG-1911, Rev 3, Washington, DC, February 2012, ADAMS Accession No. ML111890412.

Portage 2011. Portage, Inc., "Tank Farm Performance Assessment—GWSCREEN Modeling to Evaluate the Reduced Impact of the Big Lost River on Perched Water," Prepared for CH2M-WG Idaho, LLC, Contract DE-AC07-05ID14516500116.32 by Portage, Inc., Idaho Falls, ID, August 2011.

6.0 GLOSSARY

closed activity	A monitoring activity for which a key assumption made or key parameter used by the U.S. Department of Energy (DOE) in its assessment has been either substantiated or determined not to be important in meeting the performance objectives of Subpart C, "Performance Objectives," of Title 10 of the *Code of Federal Regulations* 10 CFR Part 61, "Licensing Requirements for Land Disposal of Radioactive Waste."
Factor	An assumption made or a parameter used by DOE in its performance demonstration that the U.S. Nuclear Regulatory Commission (NRC) has determined to be important through the review of a DOE waste determination, which describes its waste disposal actions and demonstrates that there is reasonable assurance that the performance objectives listed in 10 CFR Part 61, Subpart C, will be met.
highly radioactive radionuclides	Those radionuclides that contribute most significantly to risk to the public, workers, and the environment.
key monitoring area	An area that the NRC has determined, through the review of a DOE waste determination that describes its waste disposal actions, to be important to demonstrating reasonable assurance that the performance objectives listed in 10 CFR Part 61, Subpart C, will be met.
K_d (Distribution Coefficient)	A measure of the partitioning of a substance between water and a solid (e.g., cement or sediment). It describes the ability of a porous material to retain chemical constituents.
monitoring activities	NRC and State activities to monitor DOE disposal actions to assess compliance with the performance objectives listed in 10 CFR Part 61, Subpart C.

noncompliance	A conclusion that DOE disposal actions will not be in compliance with the performance objectives of 10 CFR Part 61, Subpart C, or that there is an insufficient basis to assess whether the DOE waste disposal action will result in compliance with the performance objectives.
open activity	Monitoring activity that has not been closed and for which sufficient information has not been obtained to fully assess compliance with a 10 CFR Part 61, Subpart C performance objective.
open issue	An issue that arises during monitoring activities that requires additional follow-up by the NRC staff or additional information from DOE to address questions that the NRC staff has raised regarding DOE disposal actions. Items raised to the level of becoming an open issue are typically of high risk-significance.
open-noncompliant activity	An ongoing monitoring activity that has provided evidence that the performance objectives of 10 CFR Part 61, Subpart C, are currently not being met or will not be met in the future or for which insufficient technical bases have been provided to determine that the performance objectives will be met.
operations	The timeframe during which DOE carries out its waste disposal actions through the end of the institutional control period. For the purpose of this plan, DOE actions involving waste disposal are considered to include performance assessment development (analytical modeling), waste removal, grouting, stabilization, observation, maintenance, or other similar activities.
performance assessment	A type of systematic risk analysis that addresses (1) what can happen, (2) how likely it is to happen, (3) what the resulting impacts are, and (4) how these impacts compare to specifically defined standards.

performance objectives	The 10 CFR Part 61, Subpart C, requirements for low-level waste disposal facilities that include protection of the general population from releases of radioactivity (10 CFR 61.41), protection of individuals from inadvertent intrusion (10 CFR 61.42), protection of individuals during operations (10 CFR 61.43), and stability of the disposal site after closure (10 CFR 61.44).
recommendations	As used in this report, suggestions to DOE that address ways in which DOE can make progress in closing any open activities in the staff's monitoring plan; a monitoring area for which an open issue has been previously identified and closed and for which the NRC staff suggests further action to strengthen some aspect of the DOE disposal action; and monitoring areas where no open issues or concerns were previously raised but the NRC staff recommends further improvements to DOE disposal actions.
	The NRC staff provides recommendations to DOE to provide DOE with the NRC staff's insights on one or more aspects of the disposal action being monitored. Recommendations may address (1) the ways that DOE can make progress on closing any open activities in the staff's monitoring plan; (2) a monitoring area for which an open issue has been previously identified and closed and for which the NRC staff recommends further action to strengthen some aspect of the DOE disposal action; or (3) monitoring areas for which no open issues or concerns were previously raised, but for which the NRC staff recommends further improvements to DOE disposal actions.
technical evaluation report	The NRC staff consults with DOE on the draft waste determination and prepares a technical evaluation report (TER) that documents the NRC staff's evaluation.
waste determination	DOE documentation demonstrating that a specific waste stream is not high-level waste (also known as non-high-level waste determination).

worker   DOE personnel (including contractors) who carry out operational activities at the disposal facility. For the purpose of this plan, 10 CFR Part 835, "Occupational Radiation Protection," dose limits (comparable to those in 10 CFR Part 20, "Standards for Protection against Radiation") would apply for radiation workers.

APPENDIX A: NATIONAL DEFENSE AUTHORIZATION ACT

Section 3116, Ronald W. Reagan National Defense Authorization Act for Fiscal Year 2005

SEC. 3116. DEFENSE SITE ACCELERATION COMPLETION.

(a)   IN GENERAL—Notwithstanding the provisions of the Nuclear Waste Policy Act of 1982, the requirements of section 202 of the Energy Reorganization Act of 1974, and other laws that define classes of radioactive waste, with respect to material stored at a Department of Energy site at which activities are regulated by a covered State pursuant to approved closure plans or permits issued by the State, the term "high-level radioactive waste" does not include radioactive waste resulting from the reprocessing of spent nuclear fuel that the Secretary of Energy (in this section referred to as the "Secretary"), in consultation with the Nuclear Regulatory Commission (in this section referred to as the "Commission"), determines—

(1) does not require permanent isolation in a deep geologic repository for spent fuel or high-level radioactive waste;

(2) has had highly radioactive radionuclides removed to the maximum extent practical; and

(3) (A) does not exceed concentration limits for Class C low-level waste as set out in Section 61.55 of Title 10, Code of Federal Regulations, and will be disposed of—

    (i)   in compliance with the performance objectives set out in Subpart C of Part 61 of title 10, Code of Federal Regulations; and

    (ii)   pursuant to a State-approved closure plan or State-issued permit, authority for the approval or issuance of which is conferred on the State outside of this section; or

(B) exceeds concentration limits for Class C low-level waste as set out in section 61.55 of Title 10, Code of Federal Regulations, but will be disposed of—

    (i)   in compliance with the performance objectives set out in Subpart C of Part 61 of Title 10, Code of Federal Regulations;

    (ii)   pursuant to a State-approved closure plan or State-issued permit, authority for the approval or issuance of which is conferred on the State outside of this section; and

    (iii)   pursuant to plans developed by the Secretary in consultation with the Commission.

(b)   MONITORING BY NUCLEAR REGULATORY COMMISSION

(1) The Commission shall, in coordination with the covered State, monitor disposal actions taken by the Department of Energy pursuant to Subparagraphs (A) and (B) of subsection (a)(3) for the purpose of assessing compliance with the performance objectives set out in Subpart C of Part 61 of Title 10, Code of Federal Regulations.

(2) If the Commission considers any disposal actions taken by the Department of Energy pursuant to those subparagraphs to be not in compliance with those performance objectives, the Commission shall, as soon as practicable after discovery of the noncompliant conditions, inform the Department of Energy, the covered State, and the following congressional committees:

(A) The Committee on Armed Services, the Committee on Energy and Commerce, and the Committee on Appropriations of the House of Representatives.

(B) The Committee on Armed Services, the Committee on Energy and Natural Resources, the Committee on Environment and Public Works, and the Committee on Appropriations of the Senate.

(3) For fiscal year 2005, the Secretary shall, from amounts available for defense site acceleration completion, reimburse the Commission for all expenses, including salaries, that the Commission incurs as a result of performance under subsection (a) and this subsection for fiscal year 2005. The Department of Energy and the Commission may enter into an interagency agreement that specifies the method of reimbursement. Amounts received by the Commission for performance under subsection (a) and this subsection may be retained and used for salaries and expenses associated with those activities, notwithstanding Section 3302 of Title 31, United States Code, and shall remain available until expended.

(4) For fiscal years after 2005, the Commission shall include in the budget justification materials submitted to Congress in support of the Commission budget for that fiscal year (as submitted with the budget of the President under section 1105(a) of title 31, United States Code) the amounts required, not offset by revenues, for performance under subsection (a) and this subsection.

(c) INAPPLICABILITY TO CERTAIN MATERIALS—Subsection (a) shall not apply to any material otherwise covered by that subsection that is transported from the covered State.

(d) COVERED STATES—For purposes of this section, the following States are covered States:

(1) The State of South Carolina.

(2) The State of Idaho.

(e) CONSTRUCTION

(1) Nothing in this section shall impair, alter, or modify the full implementation of any Federal Facility Agreement and Consent Order or other applicable consent decree for a Department of Energy site.

(2) Nothing in this section establishes any precedent or is binding on the State of Washington, the State of Oregon, or any other State not covered by subsection (d) for the management, storage, treatment, and disposition of radioactive and hazardous materials.

(3) Nothing in this section amends the definition of "transuranic waste" or regulations for repository disposal of transuranic waste pursuant to the Waste Isolation Pilot Plant Land Withdrawal Act or Part 191 of Title 40, Code of Federal Regulations.

(4) Nothing in this section shall be construed to affect in any way the obligations of the Department of Energy to comply with section 4306A of the Atomic Energy Defense Act (50 U.S.C. 2567).

(5) Nothing in this Section amends the West Valley Demonstration Act (42 U.S.C. 2121a note).

(f) JUDICIAL REVIEW—Judicial review shall be available in accordance with Chapter 7 of Title 5, United States Code, for the following:

(1) Any determination made by the Secretary or any other agency action taken by the Secretary pursuant to this section.

(2) Any failure of the Commission to carry out its responsibilities under Subsection (b).

APPENDIX B: MONITORING SUMMARY TABLES

Summary Tables of U.S. Nuclear Regulatory Commission Monitoring Plans

Table B-1: Monitoring at Savannah River Site Saltstone Facilities (NRC, 2007b)

10 CFR Part 61 Performance Objectives	Monitoring Area	Description	Activities		Status[2]
			Monitoring Activity Code	Type[1]	
61.41		Data review	Review information on reported inventories and concentrations in the Saltstone Disposal Facility. (SRS-SLT-41-00-01-T)	T	Open
			Review ground water monitoring data, updates to the monitoring plan, and quality assurance plans for sampling. (SRS-SLT-41-00-02-T)	T	Open
	Factor 1, Oxidation of Saltstone	The rate of waste oxidation is a key factor in the future performance of the Saltstone Disposal Facility because the release of technetium is very dependent on the extent of oxidation of the saltstone waste	Review information on vault design as it relates to oxidation. (SRS-SLT-41-01-01-T)	T	Open
			Review information on gas phase transport of oxygen within the saltstone. (SRS-SLT-41-01-02-T)	T	Open

[1]   There are two main types of monitoring activities: T=technical review activities; O=onsite observation activities.

[2]   The activities are tracked as open, open-noncompliant, or closed. The glossary defines these terms. Note that an open activity is different from an open issue.

B-1

Table B-1: Monitoring at Savannah River Site Saltstone Facilities (NRC, 2007b)

10 CFR Part 61 Performance Objectives	Monitoring Area	Description	Activities Monitoring Activity Code	Type[1]	Status[2]
61.41 (cont.)	Factor 1, Oxidation of Saltstone (cont.)	form. Realistic modeling of waste oxidation is needed to assure that the performance objectives of Title 10 of the *Code of Federal Regulations* (10 CFR) 61.41, "Protection of the General Population from Releases of Radioactivity," will be met. Adequate model support is essential to providing the technical basis for the model results.	Review field and laboratory experiments and any additional modeling of saltstone oxidation and technetium release. (SRS-SLT-41-01-03-T)	T	Open
			Review information on grout formulation and grout curing conditions. (SRS-SLT-41-01-04-O)	O	Open
			Evaluate the adequacy of the U.S. Department of Energy (DOE) program for verifying the specifications of blast furnace slag. (SRS-SLT-41-01-05-O)	O	Open
	Factor 2, Hydraulic Isolation of Saltstone	To better understand the future performance of the disposal facility, it is important to understand the mechanisms of degradation of the waste form to predict the rate of degradation, as well as the expected physical properties of the degraded waste	Review information to support the exclusion from consideration of specific saltstone degradation mechanisms. (SRS-SLT-41-02-01-T)	T	Open
			Review information on curing technique and curing time for grout and concrete. (SRS-SLT-41-02-02-T)	T	Open

Table B-1: Monitoring at Savannah River Site Saltstone Facilities (NRC, 2007b)

10 CFR Part 61 Performance Objectives	Monitoring Area	Description	Activities		Status[2]
			Monitoring Activity Code	Type[1]	
		form, such as hydraulic conductivity and diffusivity.	Review information on water condensation within the vaults. (SRS-SLT-41-02-03-T)	T	Open
			Review information on the dissolution of salts and low-solubility matrix phases within the grout. (SRS-SLT-41-02-04-T)	T	Open
			Observe vault construction and performance. (SRS-SLT-41-02-05-O)	O	Open
61.41 (cont.)	Factor 3, Model Support	Adequate model support is essential to assessing whether the saltstone disposal facility can meet the requirements of §61.41. The model support for the following items is key to confirming the performance assessment results: (1) moisture flow through fractures in the concrete and saltstone located in the vadose zone, (2) realistic modeling of waste oxidation and release of technetium, (3) the extent and frequency of fractures in saltstone and vaults that will	Review any new moisture characteristic data for concrete and saltstone. (SRS-SLT-41-03-01-T)	T	Open
			Review available information on the rate of equilibrium of water content within the saltstone. (SRS-SLT-41-03-02-T)	T	Open
			Review any additional modeling analysis of moisture flow in the saltstone. (SRS-SLT-41-03-03-T)	T	Open
			Review DOE conceptual model for oxidation and technetium release and any support for the model. (SRS-SLT-41-03-04-T)	T	Open

Table B-1: Monitoring at Savannah River Site Saltstone Facilities (NRC, 2007b)

10 CFR Part 61 Performance Objectives	Monitoring Area	Description	Activities Monitoring Activity Code	Type[1]	Status[2]
		form over time, (4) the plugging rate of the lower drainage layer of the engineered cap, and (5) the long-term performance of the engineering cap as an infiltration barrier.	Review laboratory and field studies on concrete and saltstone cracking. (SRS-SLT-41-03-05-T)	T	Open
			Observe any experiments performed to address issues related to Factor 3. (SRS-SLT-41-03-06-O)	O	Open
61.42	Factor 4, Erosion Control Design	Implementation of an adequate erosion control design is important to ensuring that the provisions of §61.42, "Protection of Individuals from Inadvertent Intrusion," can be met. The erosion control barrier will help to maintain a thick layer of soil over the vaults, which reduces the potential for intrusion into the waste.	Evaluate technical details of the proposed closure cap. (SRS-SLT-42-04-01-T)	T	Open
			Evaluate the design of erosion control features. (SRS-SLT-42-04-02-T)	T	Open
			Evaluate updates or revisions to DOE intruder analysis. (SRS-SLT-42-04-03-T)	T	Open
61.41	Factor 5, Infiltration Barrier Perf.	The design and performance of the infiltration control system is important for ensuring that the requirements of §61.41 can be met. The release of contaminants from the saltstone to the ground water is predicted to be sensitive to the amount of infiltration.	Review experiments and field studies that simulate processes related to plugging of the drainage layer through colloidal clay migration. (SRS-SLT-41-05-01-T)	T	Open
			Review any experiments, analyses, or expert elicitation regarding the long-term performance of the infiltration barrier. (SRS-SLT-41-05-01-T)	T	Open

Table B-1: Monitoring at Savannah River Site Saltstone Facilities (NRC, 2007b)

10 CFR Part 61 Performance Objectives	Monitoring Area	Description	Activities		Status[2]
			Monitoring Activity Code	Type[1]	
	Factor 6, Feed Tank Sampling	Implementation of an adequate waste sampling plan is important to ensuring that the provisions of §61.41 and §61.42 can be met. It is necessary to confirm that the concentration of highly radioactive radionuclides (HRRs) in treated salt waste (or grout) is less than or equal to the concentration assumed in the waste determination.	Review DOE waste sampling plan and quality assurance procedures for sampling waste. (SRS-SLT-41-06-01-T)	T	Open
			Review waste sampling data for the feed tank (Tank 50). (SRS-SLT-41-06-02-T)	T	Open
			Observe waste sampling activities. (SRS-SLT-41-06-03-O)	O	Open
61.41 (cont.)	Factor 7, Tank 48 Waste form	The chemical composition of the salt waste in Tank 48 differs from the salt waste in other tanks because it contains a substantial amount of organic salts. To ensure that Tank 48 waste can be safely managed, tests are needed to measure the physical properties of the waste form made from this waste to confirm that it will provide suitable performance.	Review DOE approach for treating waste in Tank 48. (SRS-SLT-41-07-01-T)	T	Open
			Review characterization information for Tank 48. (SRS-SLT-41-07-02-T)	T	Open
			Review information on the expected physical properties of the Tank 48 waste form. (SRS-SLT-41-07-03-T)	T	Open
61.41 (cont.)	Factor 8, Removal Efficiencies	The removal efficiencies of HRRs by each of the planned salt waste treatment processes are a key	Review information on radionuclide removal efficiencies by the various treatment processes. (SRS-SLT-41-08-01-T)	T	Open

Table B-1: Monitoring at Savannah River Site Saltstone Facilities (NRC, 2007b)

10 CFR Part 61 Performance Objectives	Monitoring Area	Activities Description	Monitoring Activity Code	Type[1]	Status[2]
		factor in determining the radiological inventory disposed of in saltstone, which, in turn, is an important factor in determining that §61.41 and §61.42 can be met.	Review estimates of the amount of sludge entrained in the salt waste during the deliquification, dissolution, and adjustment process. (SRS-SLT-41-08-02-T)	T	Open
			Evaluate updates or revisions to DOE performance assessment (PA) and special analysis. (SRS-SLT-41-08-03-T)	T	Open
61.43	Radiation Protection and Environmental Protection		Review reports related to worker and general public doses. (SRS-SLT-43-RE-01-T)	T	Open
			Review air effluent data from the salt waste processing facility. (SRS-SLT-43-RE-02-T)	T	Open
			Review information on DOE quality assurance program for monitoring air emissions. (SRS-SLT-43-RE-03-T)	T	Open
			Review DOE radiation protection program. (SRS-SLT-43-RE-04-O)	O	Open
			Observe DOE process for obtaining air effluent data. (SRS-SLT-43-RE-05-O)	O	Open

Table B-1: Monitoring at Savannah River Site Saltstone Facilities (NRC, 2007b)

10 CFR Part 61 Performance Objectives	Monitoring Area	Description	Activities		Status[2]	
			Monitoring Activity Code	Type[1]		
61.44			Review DOE ground water sampling process and installation of new wells. (SRS-SLT-43-RE-06-O)		O	Open
			Observe the disposal facility for obvious signs of degeneration. (SRS-SLT-44-XX-01-O)		O	Open

Table B-2: Monitoring at Idaho National Laboratory Idaho Nuclear Technology and Engineering Center Tank Farm Facility (NRC, 2007c)

10 CFR Part 61 Performance Objectives	Monitoring Area	Description	Activities		
			Monitoring Activity Code	Type [3]	Status [4]
61.41	KMA 1, Residual Waste Sampling	DOE should sample tanks WM-187 through WM-190 after cleaning, as stated in Section 2.3 of the Draft Section 3116 Determination Idaho Nuclear Technology and Engineering Center Tank Farm Facility (DOE, 2005). After cleaning, DOE should review sampling data and analysis of tanks WM-187 through WM-190 to ensure that the inventory for these tanks is not significantly underestimated (i.e., similar or better waste retrieval will be achieved).	Review sampling and analysis plans (SAPs) and data quality assessments for tanks WM-187 through WM-190. (INL-TFF-41-01-01-T)	T	Open
			Compare post cleaning WM-182 tank inventory to post cleaning tank inventories developed for WM-187 through WM-190. (INL-TFF-41-01-02-T)	T	Open
			Compare vault WM-187 liquid sampling to vault WM-185 liquid sampling. (INL-TFF-41-01-03-T)	T	Open
			Observe post cleaning sampling of tanks WM-187 through WM-190 against the SAP. (INL-TFF-41-01-04-O)	O	Open
			Observe use of video equipment to map out waste residual depths in the cleaned tanks to estimate waste residual volumes. (INL-TFF-41-01-05-O)	O	Open

[3]   There are two main types of monitoring activities: T=technical review activities; O=onsite observation activities.

[4]   The activities are tracked as open, open-noncompliant, or closed. The glossary defines these terms. Note that an open activity is different from an open issue.

Table B-2: Monitoring at Idaho National Laboratory Idaho Nuclear Technology and Engineering Center Tank Farm Facility (NRC, 2007c)

10 CFR Part 61 Performance Objectives	Monitoring Area	Description	Activities Monitoring Activity Code	Type[3]	Status[4]
61.42	KMA 1, Residual Waste Sampling (cont.)		Compare post cleaning WM-182 tank inventory to the post cleaning tank inventories developed for WM-187 through WM-190. (INL-TFF-42-01-06-T)	T	Open
61.41	KMA 2, Grout Formulation and Perf.	The final grout formulation used to stabilize the Idaho Nuclear Technology and Engineering Center (INTEC) Tank Farm Facility (TFF) waste should be consistent with design specifications, or significant deviations should be evaluated to ensure that they will not negatively impact the expected performance of the grout. The reducing capacity of the tank grout is important to mitigating the release of technetium-99. Short-term performance of as-emplaced grout should be similar to or better than that assumed in the Performance Assessment (PA) release modeling, or significant deviations should be evaluated to determine their significance	Determine whether the vendor-supplied slag has sufficient sulfide content to maintain reducing conditions in the tank grout. (INL-TFF-41-02-01-T)	T	Open
			Determine whether slag storage is sufficient to maintain the quality and chemical reactivity of the slag. (INL-TFF-41-02-02-T)	T	Closed
			Assess the short-term performance of the as-emplaced grout. (INL-TFF-41-02-03-T)	T	Open
			Evaluate the final grout formulation for consistency with design specifications. (INL-TFF-41-02-04-O)	O	Open

Table B-2: Monitoring at Idaho National Laboratory Idaho Nuclear Technology and Engineering Center Tank Farm Facility (NRC, 2007c)

10 CFR Part 61 Performance Objectives	Monitoring Area	Activities			
		Description	Monitoring Activity Code	Type[3]	Status[4]
61.41 (cont.)		with respect to the conclusions in the PA and technical evaluation report (TER). The short-term performance of the grouted vault is especially important to mitigate the release of short-lived radionuclides, such as strontium-90, from the contaminated sand pads that could potentially dominate the predicted doses from the TFF within the first few hundred years.	Evaluate the risk significance of any deviations in the final grout formulation from design specifications. (INL-TFF-41-02-05-O)	O	Open
61.41 (cont.)	KMA 2, Grout Formulation and Perf. (cont.)		Evaluate the DOE program for sampling, testing, and accepting grout materials. (INL-TFF-41-02-06-O)	O	Closed
			Verify conditions of grout placement in terms of temperature and humidity. (INL-TFF-41-02-07-O)	O	Closed
61.44			Review information on grout formulation, placements, and pours. (INL-TFF-44-02-08-T)	T	Open
61.41	KMA 3, Hydrologic Uncertainty	Relevant recent and future monitoring data and modeling activities should continue to be evaluated to ensure that hydrological uncertainties that may significantly alter the conclusions in the	Evaluate and assess the risk significance of any variations in DOE PA-predicted natural attenuation of strontium-90 through the vadose zone. (INL-TFF-41-03-01-T)	T	Open

Table B-2: Monitoring at Idaho National Laboratory Idaho Nuclear Technology and Engineering Center Tank Farm Facility (NRC, 2007c)

10 CFR Part 61 Performance Objectives	Monitoring Area	Description	Activities		
			Monitoring Activity Code	Type [3]	Status [4]
		PA are addressed. If significant new information is found, it should be evaluated against the PA and TER conclusions.	Evaluate and assess the risk significance of any increased estimates of infiltration rates at the INTEC TFF above those assumed in the DOE PA. (INL-TFF-41-03-02-T)	T	Open
			Review hydrological studies and monitoring data for new and significant information related to natural attenuation at the INTEC TFF. (INL-TFF-41-03-03-T)	T	Open
61.43	KMA 4, Monitoring during Operations	Closure and post closure operations (until the end of active institutional controls, which is 100 years) will be monitored to ensure that the performance objective in §61.43, "Protection of Individuals during Operations," can be met.	Review DOE Idaho radiation protection program to ensure that it is consistent with that described in its waste determination. (INL-TFF-43-04-01-T)	T	Open
			Review pathway analysis, environmental data collected, and DOE estimate of doses to members of the public. (INL-TFF-43-04-02-T)	T	Open
			Observe risk-significant DOE closure activities. (INL-TFF-43-04-03-O)	O	Open

Table B-2: Monitoring at Idaho National Laboratory Idaho Nuclear Technology and Engineering Center Tank Farm Facility (NRC, 2007c)

10 CFR Part 61 Performance Objectives	Monitoring Area	Description	Activities		
			Monitoring Activity Code	Type [3]	Status [4]
			Observe air sampling activities and DOE meteorological program or rely on Idaho Department of Environmental Quality (DEQ) environmental surveillance program.[5] (INL-TFF-43-04-04-O)	O	Open
61.41	KMA 5, Engineered Surface Barrier/ Infiltration Reduction	INTEC infiltration controls and the construction and maintenance of an engineered cap over the TFF under the Comprehensive Environmental Response, Compensation, and Liability program should be monitored to ensure that the PA assumptions related to infiltration and contaminant release are bounding.	Evaluate and assess the design, construction, maintenance, and as-emplaced performance of engineered barriers installed at the INTEC TFF against DOE PA assumptions regarding infiltration. (INL-TFF-41-05-01-T)	T	Open
61.41	KMA 5, Engineered Surface Barrier/ Infiltration Reduction (cont.)		Remain cognizant of any changes to the preliminary design of the infiltration-reducing cap. (INL-TFF-41-05-02-O)	O	Open
			Observe maintenance activities of the cap. (INL-TFF-41-05-03-O)	O	Open

5 As noted in the body of the report, the U.S. Nuclear Regulatory Commission (NRC) relies on the Idaho DEQ environmental surveillance program for this monitoring activity.

Table B-2: Monitoring at Idaho National Laboratory Idaho Nuclear Technology and Engineering Center Tank Farm Facility (NRC, 2007c)

10 CFR Part 61 Performance Objectives	Monitoring Area	Description	Activities		
			Monitoring Activity Code	Type [3]	Status [4]
61.41	Update Perf. Assessment	DOE Order 435.1, "Radioactive Waste Management," requires that the DOE PA be reviewed and revised when there are changes in waste form or containers, radionuclide inventories, facility design or operation, or closure concepts or there is an improved understanding of facility performance.	Review any revisions and updates to the DOE PA model to assess the impact of changes on conclusions regarding compliance with the performance objectives. (INL-TFF-41-PA-01-T)	T	Open
61.41	Environmental Review and Environmental Sampling	Routine review of environmental monitoring data (or studies), and associated sampling	Review analytical data on perched and saturated ground water at the INTEC TFF. (INL-TFF-41-RE-01-T)	T	Open
			Review hydrological studies relevant to flow and transport at the INTEC TFF. (INL-TFF-41-RE-02-T)	T	Open
61.41 and 61.43	Environmental Review and Environmental		Observe the installation of monitoring wells and instrumentation. (INL-TFF-41-RE-03-O)	O	Open

Table B-2: Monitoring at Idaho National Laboratory Idaho Nuclear Technology and Engineering Center Tank Farm Facility (NRC, 2007c)

10 CFR Part 61 Performance Objectives	Monitoring Area	Activities			
		Description	Monitoring Activity Code	Type [3]	Status [4]
	Sampling (cont.)		Observe sampling activities. or Rely on Idaho DEQ oversight program. [6] (INL–TFF–41–RE–04–O)	O	Open
61.44	N/A		Observe signs of system failure. (INL–TFF–44–XX–01–O)	O	Open
			Observe system performance after extreme events. (INL–TFF–44–XX–02–O)	O	Open

B-14

---

[6] As noted in the body of the report, the NRC relies on the Idaho DEQ environmental surveillance program for this monitoring activity.

References

DOE, 2001. U.S. Department of Energy, DOE Order 435.1, "Radioactive Waste Management." Washington, DC, August 2001.

DOE, 2005. U.S. Department of Energy, "Draft Section 3116 Determination Idaho Nuclear Technology and Engineering Center Tank Farm Facility–Draft," DOE/NE–ID–11226, Rev. 0, Idaho Falls, ID, September 2005.

NRC, 2007b. U.S. Nuclear Regulatory Commission, "Nuclear Regulatory Commission Plan for Monitoring the U.S. Department of Energy Salt Waste Disposal at the Savannah River Site in Accordance with the National Defense Authorization Act for Fiscal Year 2005," Washington, DC, May 3, 2007, ADAMS Accession No. ML070730363.

NRC, 2007c. U.S. Nuclear Regulatory Commission, "Nuclear Regulatory Commission Plan for Monitoring the Disposal Actions Taken by the U.S. Department of Energy at the Idaho National Laboratory Idaho Nuclear Technology and Engineering Center Tank Farm Facility in Accordance with the National Defense Authorization Act for Fiscal Year 2005," Washington, DC, April 13, 2007, ADAMS Accession No. ML070650222.

APPENDIX C: NRC MONITORING ACTIVITIES TIMELINE

Timelines for activities at the Savannah River Site, Saltstone Facility and at the Idaho National Laboratory Tank Farm Facility

Monitoring Activities at the Saltstone Facility at the Savannah River Site

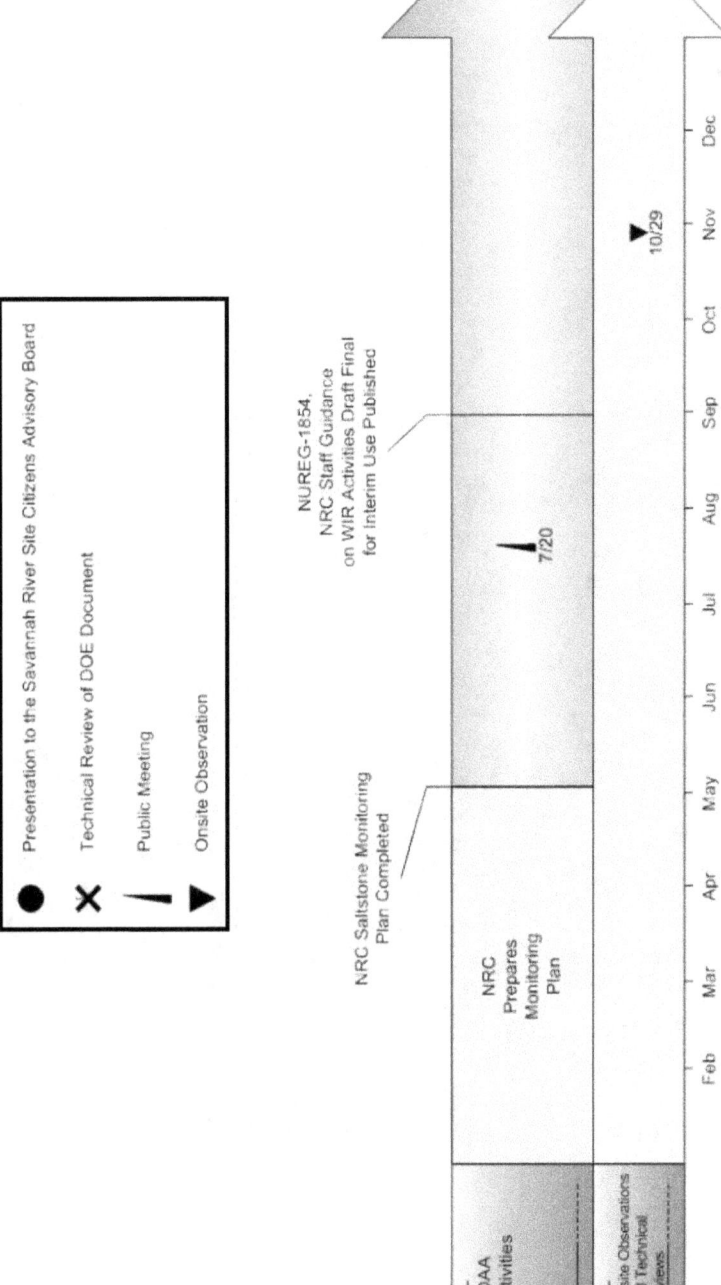

●	Presentation to the Savannah River Site Citizens Advisory Board
✕	Technical Review of DOE Document
▬	Public Meeting
▶	Onsite Observation

Figure C-1: NRC NDAA, Section 3116 Monitoring at Saltstone in 2007

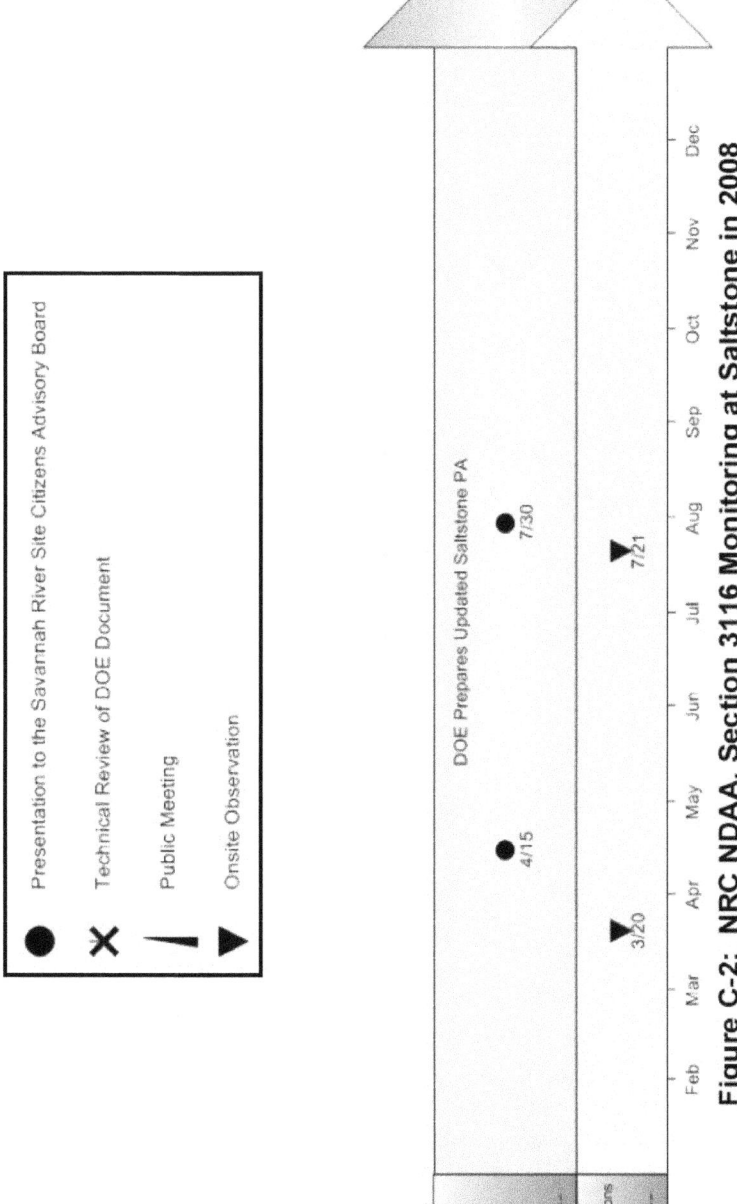

Figure C-2: NRC NDAA, Section 3116 Monitoring at Saltstone in 2008

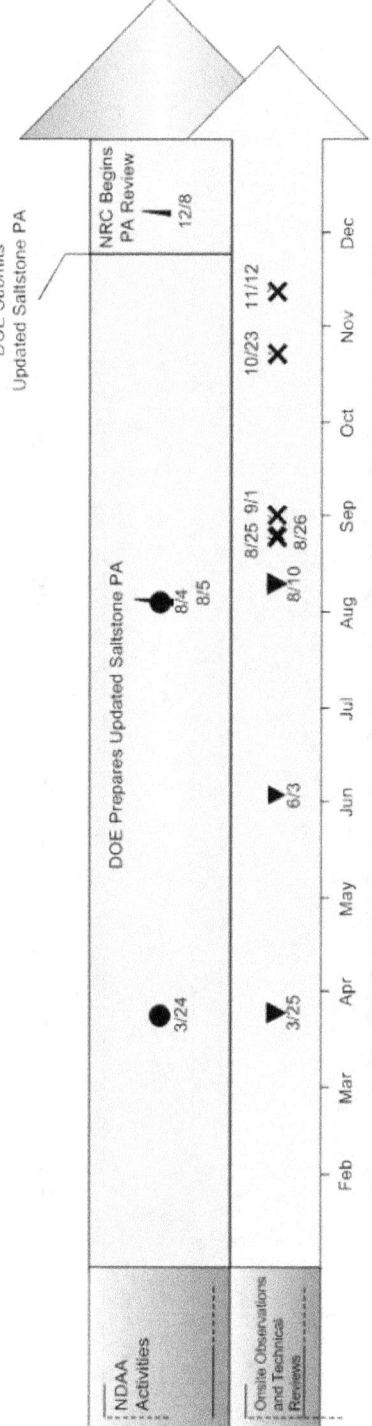

Figure C-3: NRC NDAA, Section 3116 Monitoring at Saltstone in 2009

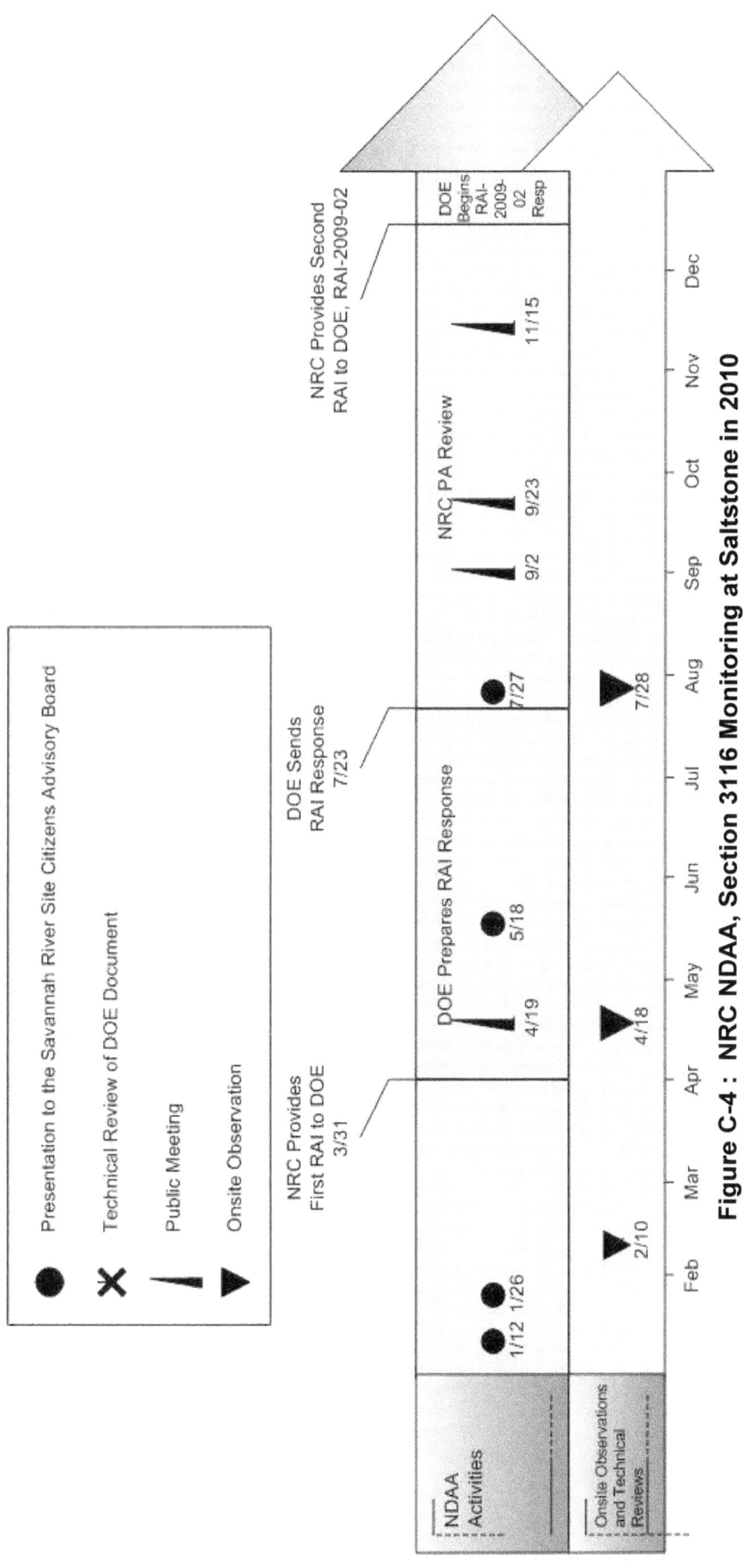

Figure C-4 : NRC NDAA, Section 3116 Monitoring at Saltstone in 2010

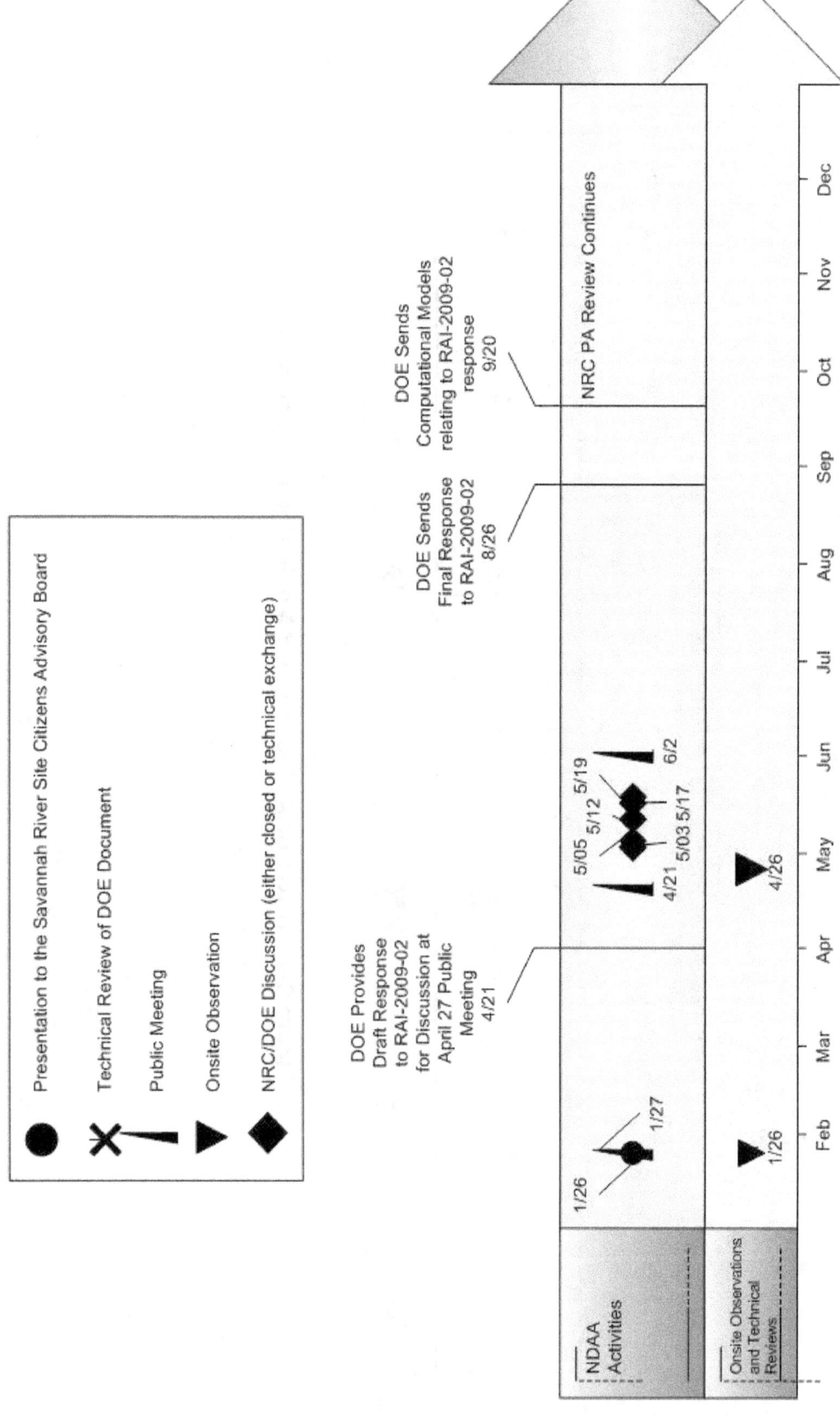

Figure C-5 : NRC NDAA, Section 3116 Monitoring at Saltstone in 2011

Monitoring Activities at the Tank Farm Facility at the Idaho National Laboratory

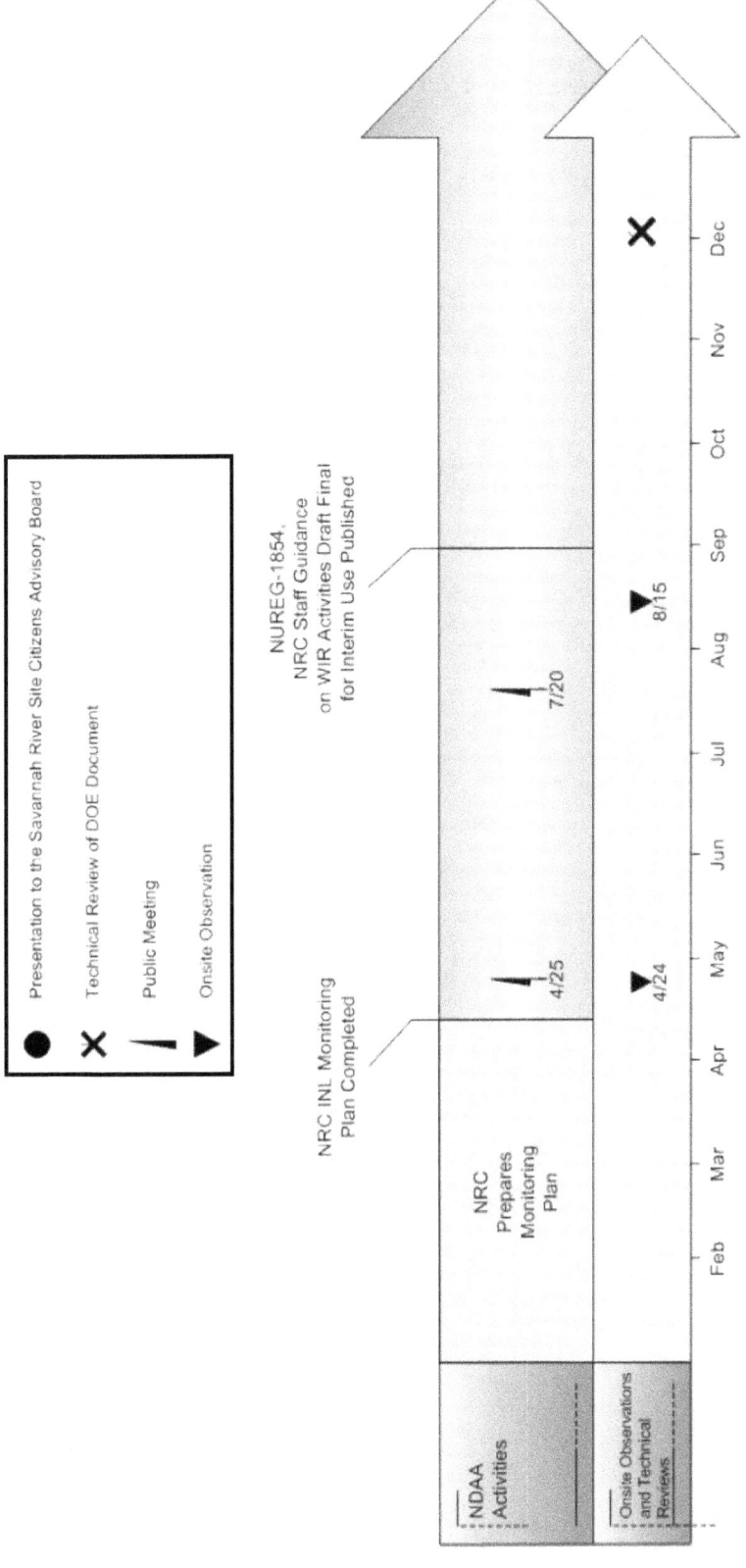

Figure C-6: NRC NDAA, Section 3116 Monitoring at INL in 2007

X Continued technical reviews for KMA 3 and KMA 4 (see NRC, 2008b for the list of technical documents reviewed)

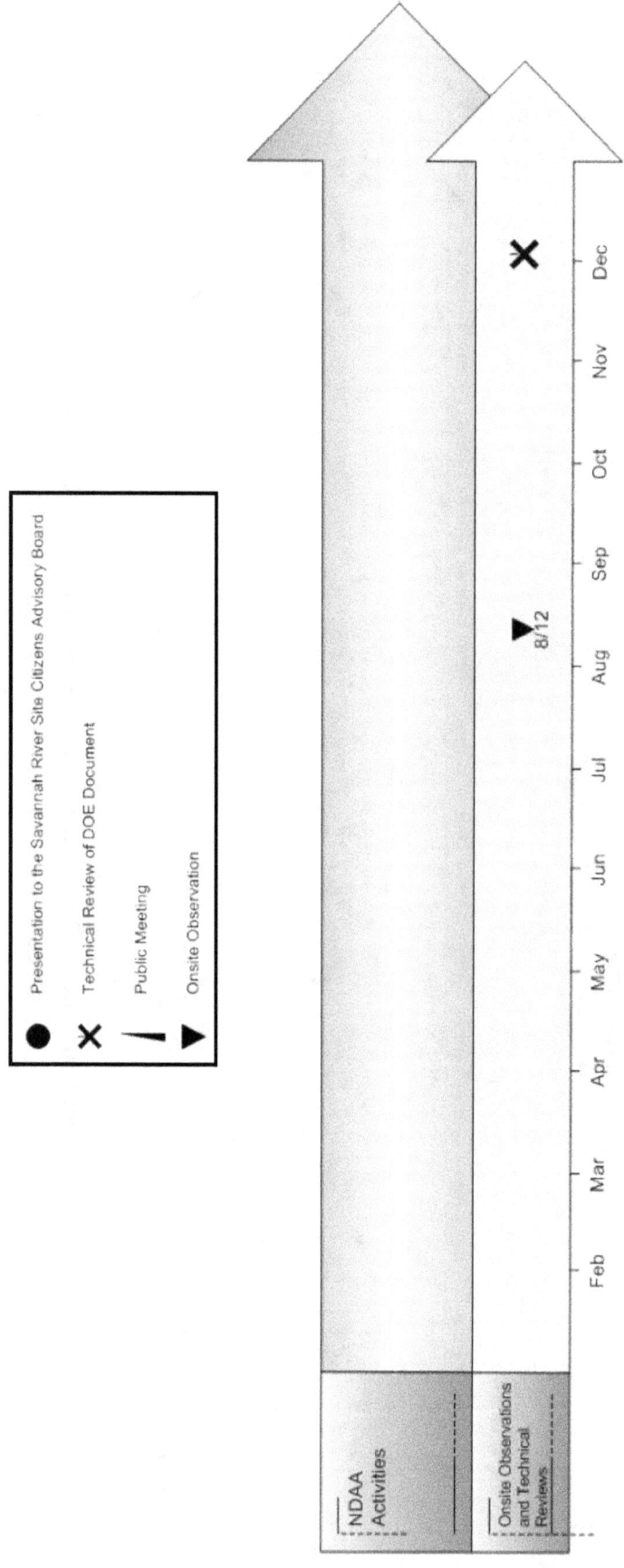

Figure C-7 : NRC NDAA, Section 3116 Monitoring at INL in 2008

X Continued technical reviews for KMA 3 and KMA 4 (see NRC, 2009b for the list of technical documents reviewed)

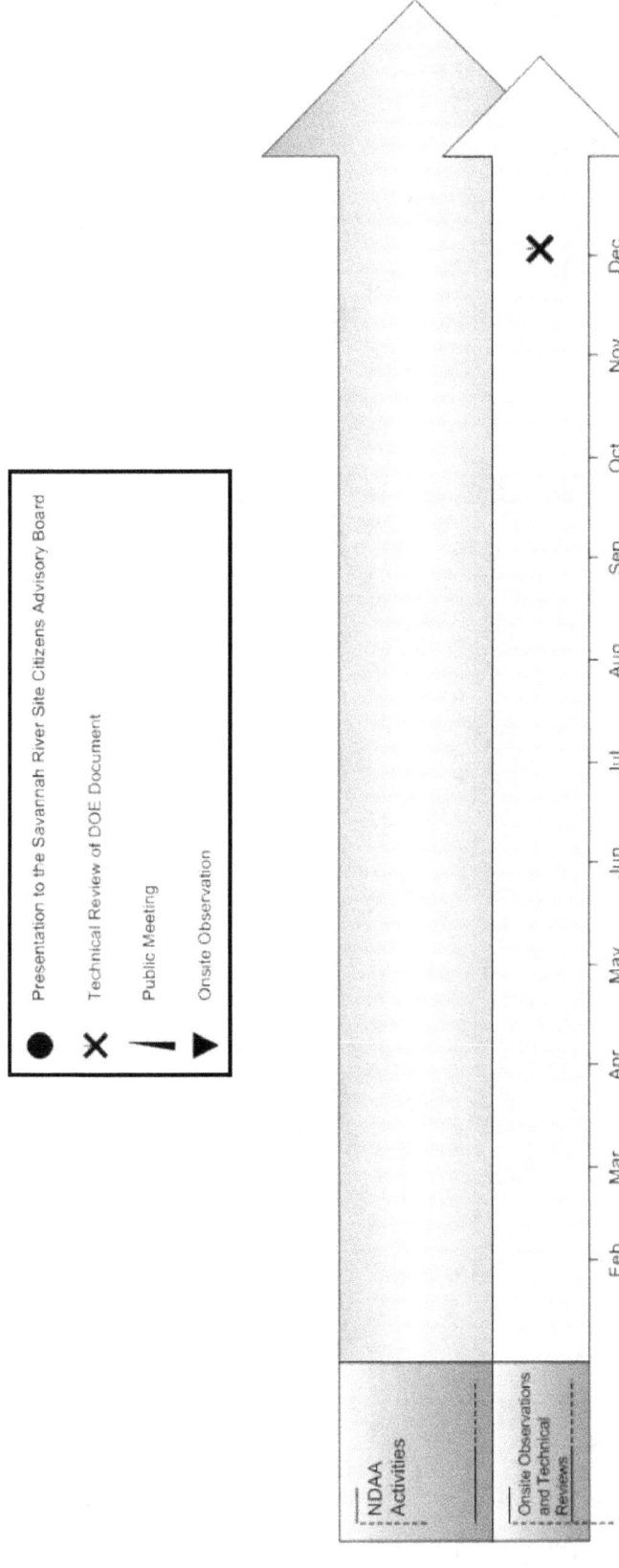

Figure C-8: NRC NDAA, Section 3116 Monitoring at INL in 2009

X Continued technical reviews for KMA 3 and KMA 4 (see NRC, 2010d for the list of technical documents reviewed)

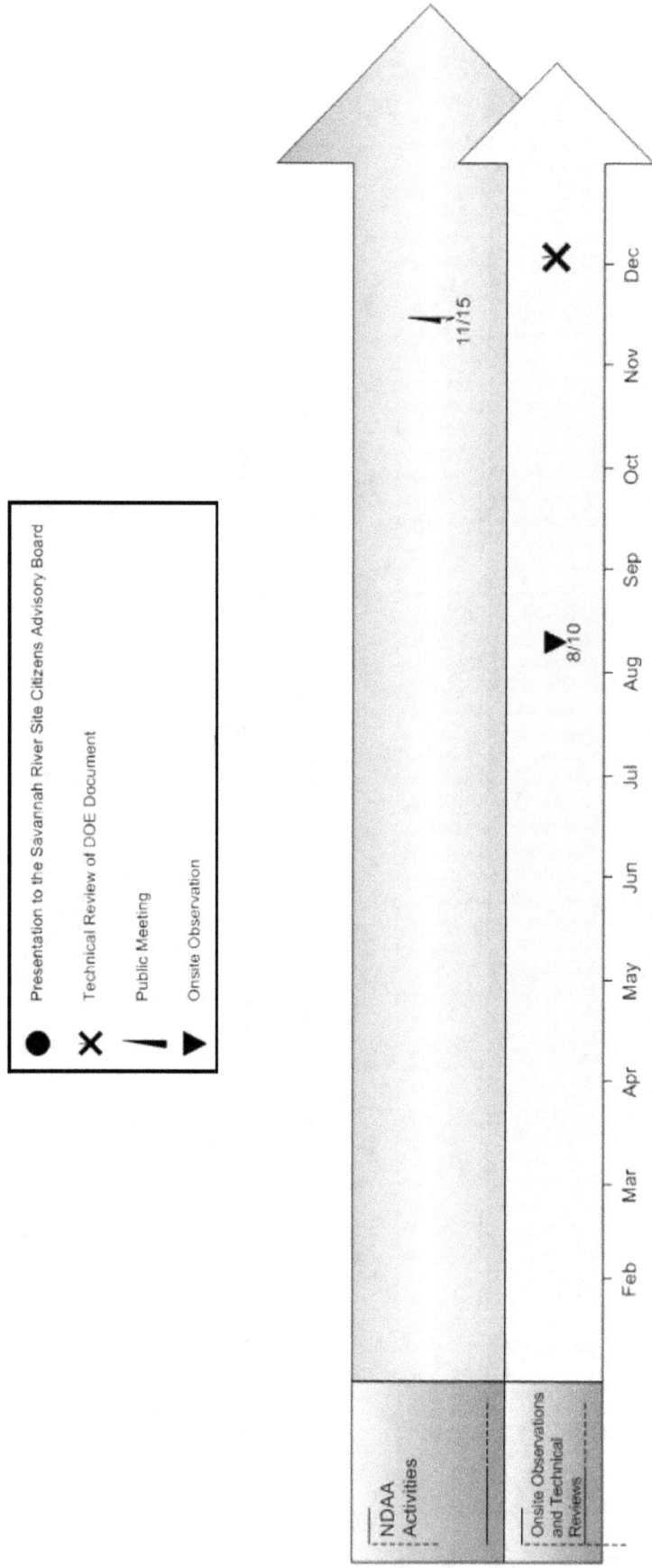

Figure C-9: NRC NDAA, Section 3116 Monitoring at INL in 2010

X Continued technical reviews for KMA 3 and KMA 4 (see NRC, 2012 for the list of technical documents reviewed)

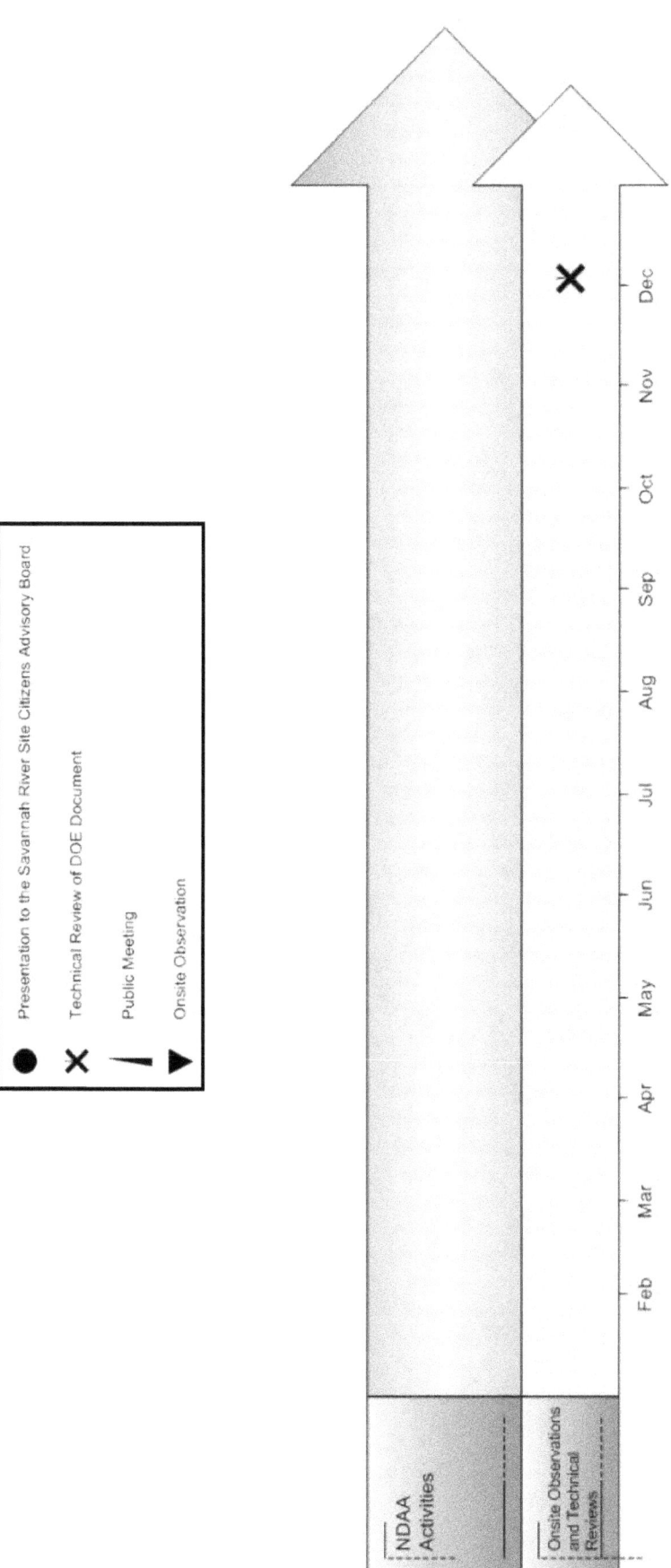

Figure C-10: NRC NDAA, Section 3116 Monitoring at INL in 2011

X Continued technical reviews for KMA 3 and KMA 4 (see Section 3.3.2 of this report for the list of technical documents reviewed)

APPENDIX D: 2011 OBSERVATION REPORTS

U.S. Nuclear Regulatory Commission Observation Reports for Calendar Year 2011

March 15, 2011

Mr. Thomas Gutmann, Director
Waste Disposition Programs Division
U.S. Department of Energy
Savannah River Operations Office
P.O. Box A
Aiken, SC 29802

SUBJECT: U.S. NUCLEAR REGULATORY COMMISSION JANUARY 27, 2011 ONSITE
             OBSERVATION REPORT FOR THE SAVANNAH RIVER SITE SALTSTONE
             FACILITY

Dear Mr. Gutmann:

The enclosed report describes the U.S. Nuclear Regulatory Commission's (NRC's) onsite observation activities on January 27, 2011, at the Savannah River Site (SRS) Saltstone Facility. This onsite observation was conducted in accordance with Section 3116 of the Ronald W. Reagan National Defense Authorization Act for Fiscal Year 2005 (Section 3116), which requires NRC to monitor disposal actions taken by the U.S. Department of Energy (DOE) for the purpose of assessing compliance with the performance objectives set out in 10 CFR Part 61, Subpart C. The activities conducted during the site visit were consistent with those described in the NRC's monitoring plan for salt waste disposal at SRS (dated May 3, 2007) and NRC's staff guidance for activities related to waste determinations (NUREG-1854, dated August 2007).

This onsite observation at SRS was focused on assessing compliance with three of the four performance objectives: (i) protection of the general population from releases of radioactivity (10 CFR 61.41), (ii) protection of individuals during operations (10 CFR 61.43), and (iii) stability of the disposal site after closure (10 CFR 61.44). Meeting these performance objectives is predicated on the performance of the disposal cells within the period of compliance.

NRC continues to conclude that there is reasonable assurance that the applicable criteria of Section 3116 can be met, if key assumptions made in DOE's waste determination analyses prove to be correct. In accordance with the requirements of Section 3116 and consistent with NRC's monitoring plan for the Saltstone Disposal Facility (SDF), NRC will continue to monitor DOE's disposal actions at SRS. The monitoring activities are expected to be an iterative process. Presently, three issues previously identified by the staff remain open: (1) the hydraulic and chemical properties of the saltstone grout, (2) the variability of saltstone from batch to batch, and (3) the reduction and retention of Technetium-99 within the saltstone waste form. Further onsite observation visits and technical reviews may be necessary in order to obtain the information needed to close all of the current open issues, as well as other issues that may be opened in the future. No discussions directly related to the three Open Issues took place during this observation.

T. Gutmann   2

If you have any questions or need additional information regarding this report, please contact
Nishka Devaser of my staff at (301) 415-5196.

                              Sincerely,

                              /RA by D. Diaz-Toro Acting for/

                              Andrew Persinko, Deputy Director
                              Environmental Protection
                                and Performance Assessment Directorate
                              Division of Waste Management
                                and Environmental Protection
                              Office of Federal and State Materials
                                and Environmental Management Programs

Enclosure:
NRC Observation Report

cc w /enclosure:
S. Wilson
Federal Facilities Liaison
Environmental Quality Control Administration
South Carolina Department of Health
  and Environmental Control
2600 Bull Street
Columbia, SC 29201-1708

U.S. NUCLEAR REGULATORY COMMISSION JANUARY 27, 2011 ONSITE OBSERVATION REPORT FOR THE SAVANNAH RIVER SITE SALTSTONE FACILITY

EXECUTIVE SUMMARY:

The U.S. Nuclear Regulatory Commission (NRC) staff conducted its tenth onsite observation visit, Observation 2011-01, to the Saltstone Facility at the Savannah River Site (SRS) on January 27, 2011. The purpose of this visit was to focus on compliance with three of the four performance objectives: (i) protection of the general population from releases of radioactivity (10 CFR 61.41), (ii) protection of individuals during operations (10 CFR 61.43), and (iii) stability of the disposal site after closure (10 CFR 61.44), by observing Vault 4 integrity and discussing saltstone production operations. This report provides a description of NRC onsite observation activities and identifies NRC observations made during the visit. Based on the results of the visit, the NRC continues to have reasonable assurance that the performance objectives of 10 CFR 61 can be met in the areas reviewed, as long as key assumptions made in the U.S. Department of Energy's (DOE's) waste determination analysis prove to be correct.

There are no new open issues resulting from this observation. The NRC staff received documentation pertaining to operations at the Saltstone Production Facility and pertaining to both the inventory of Vault 4 and the details about Cell A. Each of the documents received by the NRC staff during the observation are accessible via NRC's document repository, the Agencywide Documents Access and Management System (ADAMS), via the package accession number ML110670458.

A summary of the staff's observations and conclusions is provided below:

Vault 4 Integrity:

- DOE provided a tour of Vault 4 which included observations of the wall of Cell H and a video of some identified seepage spots on the south wall of Cell F. A representative of the DOE contractor, Savannah River Remediation (SRR) provided answers to staffs' questions during the tour and the video pertaining to the process of identifying seepage spots along the wall and the subsequent mitigative actions.

Saltstone Production Facility Operations:

- DOE provided an overview of saltstone production operations in calendar year 2010, which included details of operation (e.g., number of operating days, disposal volume, unusual work stoppages).

Enclosure

1.0   BACKGROUND:

Section 3116 of the National Defense Authorization Act for Fiscal Year 2005 (Section 3116) authorizes DOE, in consultation with the NRC, to determine that certain radioactive waste related to the reprocessing of spent nuclear fuel is not high-level waste, provided certain criteria are met. Section 3116 also requires NRC to monitor DOE disposal actions to assess compliance with the performance objectives in 10 CFR Part 61, Subpart C.

On March 31, 2005, DOE submitted a "Draft Section 3116 Determination Salt Waste Disposal Savannah River Site" to demonstrate compliance with the Section 3116 criteria including demonstration of compliance with the performance objectives in 10 CFR Part 61, Subpart C (DOE, 2005a). In its consultation role, the NRC staff reviewed the draft waste determination and concluded that there was reasonable assurance that the applicable criteria of Section 3116 could be met, provided certain assumptions made in DOE's analyses are verified via monitoring. NRC documented the results of its review in a Technical Evaluation Report issued in December 2005 (NRC, 2005). DOE issued a final waste determination in January 2006 taking into consideration the assumptions, conclusions, and recommendations documented in NRC's Technical Evaluation Report (DOE, 2006).

To carry out its monitoring responsibility under Section 3116, NRC plans to perform three types of activities: (i) technical reviews, (ii) onsite observations, and (iii) data reviews in coordination with the State of South Carolina site regulator, South Carolina Department of Health and Environmental Control (SC DHEC). These activities will focus on key assumptions – called "factors" – identified in the NRC monitoring plan for salt waste disposal at SRS (NRC, 2007). Technical reviews generally will focus on obtaining additional model support for assumptions DOE made in its PA that are considered important to DOE's compliance demonstration. Onsite observations generally will be performed to (i) observe the collection of data (e.g., observation of waste sampling used to generate radionuclide inventory data) and review the data to assess consistency with assumptions made in the waste determination, or (ii) observe key disposal (or closure) activities related to technical review areas (e.g., slag and other material storage, grout formulation and preparation, and grout placements). Data reviews will supplement technical reviews by focusing on monitoring data that may also indicate future system performance or by reviewing records or reports that can be used to directly assess compliance with performance objectives.

2.0   NRC ONSITE OBSERVATION ACTIVITIES:

The observation began with a short briefing presented by the DOE contractor, Savannah River Remediation (SRR) and attended by representatives from DOE, NRC, SC DHEC, and SRR. The briefing consisted of going through the observation agenda and reviewing standard safety considerations at the facility in preparation of a facility tour. After the briefing, Saltstone Production Facility (SPF) staff (employees of SRR) took the group on a tour of Vault 4 which consisted of observing the exterior wall of an empty vault cell, Cell H. SRR staff then moved the group into a conference room to discuss the operations at the SPF and to watch a short video of a seepage spot on Cell F.

2.1   SALTSTONE VAULT 4 WALL INTEGRITY:

2.1.1 Observation Scope:

The observation of DOE saltstone disposal operations pertains to Factor 1 – "Oxidation of Saltstone" and Factor 2 – "Hydraulic Isolation of Saltstone" identified in the NRC monitoring plan for the SRS SPF and SDF (NRC, 2007). The concrete vaults of the SDF are assumed to provide secondary containment for saltstone as well as limit wasteform exposure to aggressive chemical conditions. The objectives of the monitoring visit were to observe Vault 4 walls with respect to wasteform isolation, stability in the local environment, as well as gaining an understanding of the process SRR uses to identify seepage spots on the cell wall and the subsequent mitigative actions. Verifying the integrity of the Vault 4 walls is important to assessing the vaults ability to maintain hydraulic isolation of the saltstone waste form which relates directly to ensuring compliance with 10 CFR 61.41, "*protection of the general population from releases of radioactivity*".

Section 3.1.3, "Hydraulic Isolation of Saltstone," of the May 2007 monitoring plan (NRC, 2007) provides the basis for the staff's intended review areas.

2.1.2 Observation Results:

The staff observed the exterior wall of the Vault 4, Cell H and SRR staff discussed earlier mitigative actions to limit the release of radiologically-contaminated water from the cells during disposal operations. Seepage has occurred at imperfections in the vault walls as liquid builds up in the gap between the saltstone and vault wall. DOE has applied sealant coatings, a rain shield, certified huts, and a drip pan on the exterior of the vault cells to reduce seepage of liquid to the environment.

The vaults are intended to provide secondary containment for the radioactive saltstone wasteform. SRR staff stated that Vault 4 was not designed to be watertight and the 2009 PA assumes a very high hydraulic conductivity of 0.17 cm/s. Although this value likely bounds the potential range of hydraulic conductivities of the fractured walls of Vault 4, the moisture characteristic curve implemented in the PA for fractured concrete significantly reduces the modeled flow rate through the walls. It is not clear that the flow through the walls of Vault 4, as modeled and assumed in the 2009 PA, is consistent with observations of seepage. NRC and DOE staff will further address this issue in an upcoming observation visit.

NRC staff inquired about the integrity of the roofs of the Vault 4 cells as this provides a degree of hydraulic isolation to the wasteform. SRR staff indicated that there are active efforts to reduce the infiltration of rainwater into the cells. NRC staff requested any documentation of repair work to the roofs of Vault 4 to ensure that the assumptions in the PA regarding the hydraulic properties of the roof are consistent with ongoing observations. DOE supplied images of the repair work to the roof of Cell A, Vault 4 (ML110620217).

In addition to the tour, SRR staff presented a video of a seepage spot on the Vault 4, Cell F wall and discussed the use of disposal pads to mitigate the releases. SRR staff stated the plan is to dispose of the pads in Vault 1 or E Area. Toxicity Characteristic Leaching Procedure tests are conducted on the pads but a radionuclide-specific characterization of the pads has not been conducted. Characterization of the radionuclides that are released may provide insight into the stability of the saltstone wasteform (e.g., whether or not Tc-99 is retained in the wasteform).

2.1.3 Conclusions and Follow-up Actions:

No issues or concerns were identified during the observation of Vault 4. With respect to the Vault 4 seepage, the corrective actions taken by DOE should be effective at significantly reducing or eliminating contamination from the vault from reaching the environment in the short term (NRC, 2008). NRC staff requested any documentation regarding repair work to the roofs of all Vault 4 cells and maintains an interest in the disposal and characterization of the absorbent pads. Based on the discussion that took place during the observation, the NRC continues to have reasonable assurance that the 10 CFR Part 61 performance objectives can be met.

2.2   SALTSTONE PRODUCTION FACILITY OPERATIONS:

2.2.1 Observation Scope:

The staff's interest in discussing operations at the SPF is to ensure that the production of saltstone grout at the SDF is consistent with the assumptions made in the 2009 PA. Verifying the suitability of the saltstone production process is important to assessing the sites' radiation protection program which relates directly to ensuring compliance with 10 CFR 61.43, "protection of individuals during operations."

Section 5.2.1, "Radiation Protection Program," of the May 2007 monitoring plan (NRC, 2007) provides the basis for the staff's intended review areas.

2.2.2 Observation Results:

The NRC staff was not able to observe the saltstone grout in operation during the observation. In lieu of observing active operations, SRR staff provided a presentation explaining the current inventory being disposed of onsite, a short description of 2010 operational parameters (e.g., number of operating days, aggregate disposed inventory, quarterly run-rate), and an assessment of atypical operational parameters (e.g., unusual work stoppages, abnormal worker exposure).

In response to the NRC's request, DOE provided a chart with the details of Saltstone production during Calendar Year 2010 (ADAMS Acc No. ML110620205). The chart expresses cumulative production of Saltstone and identifies any interruptions in production during the year. Of note, the staff learned that approximately 2,630 kiloliters (694,000 gallons) and 1,481 TBq (40 kCi) of salt solution were disposed of in 2010 and that Vault 4 is expected to be at capacity sometime in early 2012. DOE is planning to dispose of 2 million gallons during 2011 and "several hundred thousand" gallons more in the beginning of 2012.

SRR staff stated the disposal of thorium-230 at the SDF is significantly below the assumed activity in the 2009 PA due to a very conservative estimation of Th-230 inventory in the PA. The 2009 PA assumes 7.5 Ci of Th-230 to be disposed of in Vault 4 and to date, DOE has disposed

5

of 0.028 Ci. NRC staff indicated that the inventory of Th-230 is not currently on DOE's website and requested the disposal history of this radionuclide into Vault 4. In addition, NRC staff inquired whether the predicted disposal of Tc-99 into Vault 4 is consistent with the assumed activity in the 2009 PA.

2.2.3 Conclusions and Follow-up Actions:

No issues or concerns were identified during the observation of the Saltstone Production Facility operations. NRC staff requested documentation of the Th-230 activity disposed of in Vault 4 to date and an updated prediction of the Tc-99 disposal activity for Vault 4. Documentation on the disposed Th-230 activity was provided by DOE (ML110620210). Based on the discussion that took place during the observation, the NRC continues to have reasonable assurance that the 10 CFR part 61 performance objectives can be met.

3.0   PARTICIPANTS:

U.S. NRC
George Alexander
Nishka Devaser
Gregory Suber

U.S. DOE-SR
Chun Pang
Patricia Suggs
Armanda Watson

SRR
Ginger Dickert
Edward Selden
F. Malcolm Smith
Aaron Staub
Steve Thomas

SC DHEC
Justin Koon
Jason Shirley
Scott Simons

U.S. DOE
Linda Suttora

4.0   REFERENCES:

U.S. Department of Energy (DOE). DOE-WD-2005-001, "Basis for Section 3116 Determination Salt Waste Disposal at the Savannah River Site." Washington, DC: DOE. March 2005a.

———. CBU-PIT-2005-0146, "Saltstone Performance Objective Demonstration Document." Westinghouse Savannah River Company. June 2005b.

———. DOE-WD-2005-001, "Basis for Section 3116 Determination Salt Waste Disposal at the Savannah River Site." Washington, DC: DOE. January 2006.

U.S. Nuclear Regulatory Commission (NRC). "Technical Evaluation Report for Draft Waste Determination for Salt Waste Disposal." Letter from L. Camper to C. Anderson, DOE. December 28, 2005. (Agencywide Documents Access and Management System (ADAMS) Accession No. ML053010225).

———. "U.S. Nuclear Regulatory Commission Plan for Monitoring the U.S. Department of Energy Salt Waste Disposal at the Savannah River Site in Accordance with the National Defense Authorization Act for Fiscal Year 2005." Washington, DC: NRC. May 3, 2007. (ADAMS Accession No. ML070730363).

————. "Nuclear Regulatory Commission March 24-28, 2008 Onsite Observation Report for the Savannah River Site Saltstone Facility." June 5, 2008 (ADAMS Accession No. ML081290367)

————. "Nuclear Regulatory Commission March 25-26, 2009 Onsite Observation Report for the Savannah River Site Saltstone Facility." May 22, 2009 (ADAMS Accession No. ML091320439)

————. "U.S. Nuclear Regulatory Commission Request for Additional Information on the 2009 Performance Assessment for the Saltstone Disposal Facility at the Savannah River Site." March 31, 2010a (ADAMS Accession Number ML100820097).

————. "Nuclear Regulatory Commission April 19, 2009 Onsite Observation Report for the Savannah River Site Saltstone Facility." July 7, 2010b (ADAMS Accession No. ML101460044)

Savannah River Remediation (SRR). SRR-CWDA-2009-00017, Performance Assessment for the Saltstone Disposal Facility at the Savannah River Site, Savannah River Site, Aiken, SC, October 29, 2009 (ADAMS Accession No. ML101590008).

August 19, 2011

Mr. Thomas Gutmann, Director
Waste Disposition Programs Division
U.S. Department of Energy
Savannah River Operations Office
P.O. Box A
Aiken, SC 29802

SUBJECT: U.S. NUCLEAR REGULATORY COMMISSION APRIL 26, 2011 ONSITE
OBSERVATION REPORT FOR THE SAVANNAH RIVER SITE SALTSTONE
FACILITY

Dear Mr. Gutmann:

The enclosed report describes the U.S. Nuclear Regulatory Commission's (NRC's) onsite observation activities on April 26, 2011, at the Savannah River Site (SRS) Saltstone Facility. This onsite observation was conducted in accordance with Section 3116 of the Ronald W. Reagan National Defense Authorization Act for Fiscal Year 2005 (Section 3116), which requires NRC to monitor disposal actions taken by the U.S. Department of Energy (DOE) for the purpose of assessing compliance with the performance objectives set out in 10 CFR Part 61, Subpart C. The activities conducted during the site visit were consistent with those described in the NRC's monitoring plan for salt waste disposal at SRS (dated May 3, 2007) and NRC's staff guidance for activities related to waste determinations (NUREG-1854, dated August 2007).

This onsite observation at SRS was focused on assessing compliance with the four performance objectives: (i) protection of the general population from releases of radioactivity (§61.41), (ii) protection of individuals from inadvertent intrusion (§61.42), (iii) protection of individuals during operations (§61.43), and (iv) stability of the disposal site after closure (§61.44). Meeting these performance objectives is predicated on the performance of the disposal cells within the period of compliance.

NRC continues to conclude that there is reasonable assurance that the applicable criteria of Section 3116 can be met, if key assumptions made in DOE's waste determination analyses prove to be correct. In accordance with the requirements of Section 3116 and consistent with NRC's monitoring plan for the Saltstone Disposal Facility (SDF), NRC will continue to monitor DOE's disposal actions at SRS. Presently, three issues previously identified by the staff remain open: (1) the hydraulic and chemical properties of the saltstone grout, (2) the variability of saltstone from batch to batch, and (3) the reduction and retention of Technetium-99 within the saltstone waste form. Further onsite observation visits and technical reviews may be necessary in order to obtain the information needed to close all of the current open issues, as well as other issues that may be opened in the future.

T. Gutmann   2

If you have any questions or need additional information regarding this report, please contact Nishka Devaser of my staff at (301) 415-5196.

                                    Sincerely,

                                    /RA/

                                    Andrew Persinko, Deputy Director
                                    Environmental Protection
                                      and Performance Assessment Directorate
                                    Division of Waste Management
                                      and Environmental Protection
                                    Office of Federal and State Materials
                                      and Environmental Management Programs

Enclosure:
NRC Observation Report

cc w /enclosure:

WIR Service List

S. Wilson
Federal Facilities Liaison
Environmental Quality Control Administration
South Carolina Department of Health
 and Environmental Control
2600 Bull Street
Columbia, SC 29201-1708

U.S. NUCLEAR REGULATORY COMMISSION APRIL 26, 2011 ONSITE OBSERVATION REPORT FOR THE SAVANNAH RIVER SITE SALTSTONE FACILITY

EXECUTIVE SUMMARY:

The U.S. Nuclear Regulatory Commission (NRC) staff conducted its eleventh onsite observation visit, Observation 2011-02, to the Saltstone Facility at the Savannah River Site (SRS) on April 26, 2011. The purpose of this visit was to focus on compliance with the performance objectives set out in 10 CFR Part 61, Subpart C: (i) protection of the general population from releases of radioactivity (§61.41), (ii) protection of individuals from inadvertent intrusion (§61.42), (iii) protection of individuals during operations (§61.43), and (iv) stability of the disposal site after closure (§61.44). To accomplish these goals, NRC staff discussed testing of saltstone properties, Vault 4 inventory, disposal unit construction, and recent research on Tc reduction and oxidation in saltstone performed by SRS. This report provides a description of NRC onsite observation activities and identifies NRC observations made during the visit. Based on the results of the visit, the NRC continues to have reasonable assurance that the performance objectives of 10 CFR 61 can be met in the areas reviewed, as long as key assumptions made in the U.S. Department of Energy's (DOE's) waste determination analysis prove to be correct.

There are no new open issues resulting from this observation. The NRC staff received documentation during the observation that pertained to the observation activities scheduled for this onsite observation. Each of the documents received by the NRC staff during the observation are accessible via NRC's document repository, the Agencywide Documents Access and Management System (ADAMS), via the package accession number ML111310169.

A summary of the staff's observations and conclusions is provided below:

Technical Discussion – Saltstone Radionuclide Inventory:

- DOE contractor staff presented a reevaluation of the inventory of I-129 currently disposed of in Vault 4 based on sample results. The NRC staff believes that the method used for this reevaluation seems reasonable. NRC staff will continue to monitor DOE's inventory tracking methodology.

- DOE provided NRC staff with an updated inventory document containing the inventory of radionuclides disposed in the Saltstone Disposal Facility as of 9/30/10.

Technical Discussion – New Research on Long-Term Testing Waste Oxidation and Technetium Release:

- DOE proposed closing Open Issue 2009-1 due to successful results of recent research. NRC expressed concern with some details of the new research.

- NRC staff concerns with new research were sufficient such that Open Issue 2009-1 remains open.

<u>Discussion of Disposal Unit 2 Construction:</u>

- DOE discussed their strategy for repairing the hydraulic leaks and provided documentation describing the design changes made to the new disposal cells.

- The NRC staff will continue to monitor the construction of the new disposal cells.

<u>Follow-up Discussion – Topics from Previous Observations:</u>

- In addition to the specific technical discussions that took place during the observation, additional topics were discussed to follow-up on discussions that took place during previous observation activities. A detailed list and description of these topics can be found later in this report. Several items were closed from this discussion; however, some items discussed remain open and will require further discussion.

1.0   BACKGROUND:

Section 3116 of the National Defense Authorization Act for Fiscal Year 2005 (Section 3116) authorizes DOE, in consultation with the NRC, to determine that certain radioactive waste related to the reprocessing of spent nuclear fuel is not high-level waste, provided certain criteria are met. Section 3116 also requires NRC to monitor DOE disposal actions to assess compliance with the performance objectives in 10 CFR Part 61, Subpart C.

On March 31, 2005, DOE submitted a "Draft Section 3116 Determination Salt Waste Disposal Savannah River Site" to demonstrate compliance with the Section 3116 criteria including demonstration of compliance with the performance objectives in 10 CFR Part 61, Subpart C (DOE, 2005a). In its consultation role, the NRC staff reviewed the draft waste determination and concluded that there was reasonable assurance that the applicable criteria of Section 3116 could be met, provided certain assumptions made in DOE's analyses are verified via monitoring. NRC documented the results of its review in a Technical Evaluation Report issued in December 2005 (NRC, 2005). DOE issued a final waste determination in January 2006 taking into consideration the assumptions, conclusions, and recommendations documented in NRC's Technical Evaluation Report (DOE, 2006).

To carry out its monitoring responsibility under Section 3116, NRC performs three types of activities: (i) technical reviews, (ii) onsite observations, and (iii) data reviews in coordination with the State of South Carolina site regulator, South Carolina Department of Health and Environmental Control (SC DHEC). These activities focus on key assumptions – called "factors" – identified in the NRC monitoring plan for salt waste disposal at SRS (NRC, 2007). Technical reviews generally focus on obtaining additional model support for assumptions DOE made in its Performance Assessment (PA) that are considered important to DOE's compliance demonstration. Onsite observations generally are performed to (i) observe the collection of data (e.g., observation of waste sampling used to generate radionuclide inventory data) and review the data to assess consistency with assumptions made in the waste determination, or (ii) observe key disposal (or closure) activities related to technical review areas (e.g., slag and other material storage, grout formulation and preparation, and grout placements). Data reviews supplement technical reviews by focusing on monitoring data that may also indicate future system performance or by reviewing records or reports that can be used to directly assess compliance with performance objectives.

2.0   NRC ONSITE OBSERVATION ACTIVITIES:

The observation began with a short briefing on the observation agenda and site safety procedures presented by the DOE contractor, Savannah River Remediation (SRR) and attended by representatives from DOE, NRC, Savannah River National Laboratory (SRNL), and SRR. The observation continued with a discussion between NRC, DOE, and associated DOE contractor staff regarding the inventory of Vault 4, the new technetium oxidation research, and various follow-up discussions from previous observations. Sections below contain detailed accounts of these discussions.

2.1   Technical Discussion – Saltstone Radionuclide Inventory

2.1.1   Observation Scope:

As noted in Section 3.1.1.1, "Data Reviews – Radioactive Inventory" of the May 2007 monitoring plan, it is important for NRC staff to verify the radioactive inventory disposed of at the Saltstone Disposal Facility because the inventory is an important factor in the compliance with the performance objective identified in §61.41, "*Protection of the General Population from Releases of Radioactivity*" and §61.42 "*Protection of Individuals from Inadvertent Intrusion.*"

2.1.2   Observation Results:

During the observation, DOE provided a presentation for discussion on Vault 4 Inventory Reporting (ML111310182). The discussion focused on the calculated inventory of I-129 and the inventory of I-129 assumed for Vault 4 in the 2009 Performance Assessment (PA).

NRC staff sent an email containing three questions related to inventory to DOE prior to this April 2011 onsite observation. The questions and DOE's responses are listed below.

NRC Inventory Question	Discussion Points
NRC staff asked DOE to provide the inventory of each radionuclide disposed of in Vault 4 since March of 2009 (i.e., since X-CLC-Z-00027 [ML102160640] was published).	DOE provided the NRC staff with the document X-CLC-Z-00034, "*Inventory Determination of PODD/SA Radionuclides in the Saltstone Disposal Facility Through 9/30/10*" (ML111310276), which provided the requested information. NRC staff noted that they would appreciate receiving a document with this information yearly when it is produced. This issue was resolved during the observation.

4

NRC Inventory Question	Discussion Points
NRC staff asked DOE to provide the method used to estimate the predicted Th-230 in the 2009 PA and the method currently being used to track the inventory of Th-230 disposed of in Vault 4.	This question, which is related to RAI IN-5, was discussed in detail during the NRC/DOE public meeting on April 27, 2011 (ML111950042). This issue was not resolved during the observation, but will be addressed by DOE in its response to the NRC staff's second RAI.
NRC staff asked DOE to indicate how the current inventory in Vault 4 compares to the assumed inventory in the 2009 PA. The NRC staff noted that, based on X-CLC-Z-00027 (ML102160640) and the quarterly monitoring reports, it appears that the I-129 disposed of in Vault 4 to date exceeds the inventory predicted in the revised PA.	NRC staff was concerned that the estimated inventory in the 2009 PA did not bound the actual inventory, and therefore, the PA might not adequately capture dose. NRC staff noted that the inventory of I-129 in the 2009 PA (0.28 Ci) was less than the reported inventory already disposed in Vault 4 (0.3 Ci), based on the website quarterly reports. I-129 is a dose significant radionuclide that was identified as a highly radioactive radionuclide (HRR) in the 2005 Waste Determination. In addition, Vault 4 has not been completely filled yet, so it is expected that the final inventory in this vault will be higher.

DOE indicated the I-129 inventory in Vault 4 does not exceed the inventory predicted in the revised PA (SRR-CWDA-2011-00070) (ML111310182) based on a reevaluation of the inventory of I-129 disposed of to date in Vault 4. DOE noted that the preliminary concentrations of I-129 reported in the quarterly reports were based on estimates determined using the Tank 50 material balance and were not based directly on sample results. DOE believes that this approach can lead to lead to an overestimation of the radionuclide inventories. When DOE compared the concentration of I-129 reported in the quarterly reports to the Tank 50 sample results, they found that the concentration of I-129 reported in the quarterly sample reports was significantly higher than the concentration measured in the Tank 50 sample during 2009. During this time period, Tank 50 had a relatively low volume of liquid in it, and the inputs into Tank 50 primarily consisted of inputs from H-canyon. The H-canyon waste stream contains I-129 at a concentration that is below the detection limit. However, in the materials balance calculation, the concentration of I-129 in this waste stream is assumed to be at the detection limit. DOE performed a recalculation of the I-129 inventory based on the sample results and estimated that the inventory in Vault 4 was 0.16 Ci. |

D-15

NRC Inventory Question	Discussion Points
(cont.) NRC staff asked DOE to indicate how the current inventory in Vault 4 compares to the assumed inventory in the 2009 PA. The NRC staff noted that, based on X-CLC-Z-00027 (ML102160640) and the quarterly monitoring reports, it appears that the I-129 disposed of in Vault 4 to date exceeds the inventory predicted in the revised PA.	DOE therefore concluded that the final inventory of I-129 in Vault 4 will be bounded by the assumed inventory in the PA. DOE stated that the final inventories for each vault will be determined using sample results, or best available information, once the filling of the vault has been completed. The NRC staff believes that the method used in DOE's reevaluation of the inventory of I-129 in Vault 4 seems reasonable and this issue was resolved during the observation. Based on this reevaluation, the current inventory of I-129 in Vault 4 is estimated to be less than the inventory assumed in the 2009 PA. The DOE indicated that they would be providing the NRC with the final inventory on an annual basis under monitoring.

2.1.3 Conclusions and Follow-up Actions:

Questions (1) and (3) were resolved during the observation. Question (2) will be addressed in DOE's response to the NRC staff's second RAI. No additional issues or concerns were identified during the technical discussion on the radionuclide inventory of Vault 4 apart from DOE's continued effort to respond to RAI-2009-02.

2.2 Technical Discussion – New Research on Long-Term Testing Waste Oxidation and Technetium Release:

2.2.1 Observation Scope:

As noted in Section 3.1.2, "Factor 1 – Oxidation of Saltstone" of the May 2007 monitoring plan, saltstone oxidation is considered to be important to compliance with the performance objectives primarily because oxidation can lead to increased releases of technetium from the wasteform. The release of Tc-99 from the wasteform is an important factor in the compliance with the performance objective identified in §61.41, "*Protection of the General Population from Releases of Radioactivity*" and §61.42 "*Protection of Individuals from Inadvertent Intrusion.*"

2.2.2 Observation Results:

For greater detail of the discussion that took place during this part of the observation, please refer to SRR-CWDA-2011-00071 (ML111310199), the DOE document provided to the NRC during the observation.

Based on the results of recent research (SRNL-STI-2010-00667 and SRNL-STI-2010-00668) (ML111310222 and ML111310234), DOE proposed to close Open Issue 2009-1, related to the initial chemical reduction of and the K_d value for Tc-99 in saltstone. DOE measured K_d values up to ~700 mL/g for Tc to saltstone formulated with 45% slag (nominal concentration) under a nitrogen atmosphere with 2% hydrogen gas. NRC staff questioned whether results obtained in an atmosphere with 2% hydrogen are applicable to as-emplaced saltstone. In addition, the slag-free control samples had similar measured K_d values for Tc-99, which indicates that the reduction and sorption of the Tc was not caused by the slag and might have been caused by the hydrogen gas instead. DOE indicated that, because the E_H of the leachate decreased with increasing slag concentrations, they conclude that slag controlled the E_H in the reducing cementitious materials.

DOE measured less sorption (K_d of 139 mL/g) of Tc-99 onto cores of saltstone taken from Vault 4, cell E (SRNL-STI-2010-00667) (ML111310222). DOE hypothesized that the K_d value was significantly less than 1000 mL/g because 30-60 ppm oxygen present in the glove box oxidized the saltstone. NRC suggested that a complete response to the open issue would indicate whether this range of oxygen concentrations could be present in the as-emplaced saltstone environment.

2.2.3 Conclusions and Follow-up Actions:

Because of the NRC staff's concerns discussed above, Open Issue 2009-1 remains open. No additional issues or concerns were identified during the technical discussion regarding technetium release and oxidation in the saltstone wasteform.

2.3   Discussion of Disposal Unit 2 Construction:

2.3.1 Observation Scope:

The staff's interest in discussing construction activities of the new disposal cells relates to ensuring the integrity of the disposal units and identifying the potential mechanisms of contaminant release from the facility. Section 3.1.3, "Hydraulic Isolation of Saltstone," of the May 2007 monitoring plan (NRC, 2007) provides details of the basis for the staff's intended review areas.

2.3.2 Observation Results:

DOE discussed cell design changes to deal with hydraulic leaks include flush cutting anchor bolts, cold cap of type V concrete without anchor bolt, washer and nut mechanical seal, flexible coating (see also Hydrotest Results in the table below). DOE provided SRR-CWDA-2011-00082 (ML111320032 and ML111320049) which describes the design changes made to the new disposal cells.

2.3.3 Conclusions and Follow-up Actions:

The NRC staff will continue to monitor the construction of the new disposal cells and will continue to monitor the cells when they are put into operation.

2.4   Follow-up Discussion – Topics from Previous Observations:

2.4.1   Observation Scope:

The staff's interest in discussing the list of topics in this section relates to multiple sections of the May 2007 monitoring plan and also relates to all four of the §61.41 performance objectives. Topics discussed included the following:

- *Performance Assessment/Research Activity*
- *Open Issues 2007-1, 2007-2, and 2009-1*
- *Follow-up Actions*
    - *Disposal Unit 2 Water Tightness Test Quality Assurance Records*
    - *Radiological Composition of Inadvertent Transfer Material*
    - *Status of ARP/MCU Management Control Plan*
    - *Assess Impact of Anchor Bolt Penetrations in Vault 4*
    - *Develop Data Related to Impact of Scale on Formed Core Sampling Methodology*
- *Email Questions from NRC Staff*
    - *Curing Temperature*
    - *Saltstone Fracturing*
    - *Hydrotest Results*
    - *Vault 4 Floor Performance*

2.4.2   Observation Results:

During the observation, DOE provided a document, SRR-CWDA-2011-00043 (ML111310214), which contains many of the details of the discussion provided in this section of the report. This document provides good context on the topics below.

- *Performance Assessment/Research Activity*

    DOE discussed current and future SDF PA maintenance activities. The current research activities include 11 studies on parameters such as the reducing environment, dispersion coefficients, degradation mechanisms, closure cap infiltration, and hydrology/geology. The planned PA maintenance activities include degradation studies, impacts of waste oxidation, vault cracking and attendant transport, and code upgrade.

- *Open Issues 2007-1 and 2007-2*

  Open Issues 2007-1 and 2007-2: DOE described plans to continue efforts to determine the hydraulic and chemical properties of as-emplaced saltstone grout. DOE indicated it would complete analysis of existing saltstone core samples and use formed-core sampling to verify the characteristics of as-emplaced saltstone. DOE is developing an integrated sampling plan to correlate the properties of laboratory-prepared and as-emplaced saltstone samples. DOE indicated it was working to quantify variability in the dry feed and the water-to premix ratios. DOE also indicated it is working to test the hydraulic and physical properties of saltstone formed with various dry feed compositions and cure temperature profiles. Determining the impact of these variations on the performance assessment is planned future work. NRC staff indicated that the plans to address the open issues sound reasonable. These two issues remain open at this time.

- *Open Issue 2009-1*

  The discussion on Open Issue 2009-1 is described in detail in Section 2.2 above. As noted above, this issue remains open at this time.

- *Follow-up Action: Disposal Unit 2 Water Tightness Test Quality Assurance Records*

  DOE will provide NRC staff with documentation of cell design changes and hydrotesting results for review when they are available following the Operational Readiness Review. This Follow-up action remains open.

- *Follow-up Action: Radiological Composition of Inadvertent Transfer Material*

  During the July 2010 onsite observation, the NRC staff requested information on the radionuclide composition of the salt solution that was inadvertently transferred to Vault 4. DOE provided document SRR-WSE-2010-00186 (ML111780337) to the NRC on October 26, 2010 to respond to this request.

  The inadvertent transfer of approximately 1900 gallons (7192 L) of liquid salt solution to Vault 4 occurred on May 19, 2010. This inadvertent transfer was caused by valve misalignment during tests of the Salt Feed Tank agitator. Following the inadvertent transfer, drain water removal was performed to remove the salt waste. DOE estimates that less than 50 gallons (189 L) of this material remained on top of the saltstone monolith following this removal. When Saltstone Production Facility was restarted, clean grout was added for 15-20 minutes to attempt to encapsulate this remaining liquid.

  Because the radiological inventory disposed of in Vault 4 is determined as the waste is transferred from Tank 50, the inventory in the inadvertent transfer material is already accounted for in the total inventory disposed of in Vault 4. However, it is useful to understand the radiological content of this material because the inventory in the inadvertent transfer material may not be encapsulated in the grout well because it was not disposed of in the form of grout. The removal of the inadvertently transferred liquid to the maximum extent practical and the placement of clean grout when the Saltstone Production Facility was restarted likely reduced this concern.

The material in the inadvertent transfer consisted of salt waste that originated in Tank 50 plus clean cap drain water returns. A dip sample was taken of the salt waste solution remaining in the hopper at the time of the inadvertent transfer. This sample was characterized for chemical constituents, but the radiological constituents were not characterized. The sample was completely used in the chemical analyses, and no sample remains.

DOE estimated the radiological content of the material in the inadvertent transfer based on the radiological composition of the waste from Tank 50 and the estimated dilution from the clean cap drain water. The dilution was estimated based on the ratio of sodium in the dip sample to the Tank 50 sample.

NRC staff has reviewed this information and has concluded that while it would have been preferable to have actual radiological characterization data for the material in the inadvertent transfer, the approach used by DOE to estimate the radiological content of this material seems reasonable. Based on the information provided to the NRC in SRR-WSE-2010-00186 (ML111780337), the NRC staff considers this action item to be closed.

- *Follow-up Action: Status of ARP/MCU Management Control Plan*

    During the March 2009 onsite observation (ML091320439), NRC staff identified a follow-up action for DOE to inform NRC when they exit the ARP/MCU management control plan. The basis for this follow-up action was that it was the understanding of the NRC staff that the sampling data obtained under the ARP/MCU management control plan was going to be used to represent the input from ARP/MCU in the material balance for Tank 50. However, during this observation, DOE stated that this sampling information was not going to be used as part of the inventory determination in the Tank 50 material balance.

    A new follow-up action from this observation is for DOE to clarify to the NRC staff how the sampling data obtained under the ARP/MCU management control plan is used and how the inventory of radionuclides sent to Tank 50 from ARP/MCU is determined.

    Additionally, DOE raised the concern that tracking long-term items (such as the exit strategy for the ARP/MCU management control plan) as follow-up actions, might not be the most efficient mechanism. NRC staff stated that a revised monitoring plan will be developed following the completion of the TER for the revised PA. In the new monitoring plan, NRC staff will generate a list of major changes to the salt waste disposal process that they would like to be made aware of, if and when they occur. The transition from the ARP/MCU management control plan is an example of what would be included on this list, and the follow-up action for DOE to notify NRC when the management control plan is exited can be handled in this manner in the future.

    DOE stated that they have no immediate plans to cease operating under ARP/MCU Management Control Plan.

During the onsite observation, DOE staff asked the NRC staff why they were interested in the status of the ARP/MCU Management Control Plan. NRC staff stated that the basis for their interest in the status of the ARP/MCU Management Control Plan was that they believed that the sample results obtained under this plan were used for determining the inventory of material transferred from ARP/MCU to Tank 50. NRC staff was also interested in knowing when the operations under the ARP/MCU Management Control Plan are ceased because DOE had previously indicated samples would be taken less frequently once this happened.

A follow-up discussion on this topic was held by phone on 6/30/11. During this phone call, DOE and contractor staff provided additional information to the NRC staff regarding the methodology used to determine the inventory transferred to Tank 50 from ARP/MCU.

DOE contractor staff stated that the purpose of the ARP/MCU samples is to obtain information related to safety (such as criticality) and process information and that these samples are not used to develop inventory information for the Saltstone Disposal Facility. Instead, the inventory information is based on direct analytical measurements of the Tank 50 samples, and the materials balance calculations for Tank 50. The inventory assumed for the ARP/MCU feed stream in the materials balance is based on the expected characterization of the waste in the particular salt batch. This assumed characterization may vary between different salt batches. DOE contractor staff stated that this assumed characterization is typically an upper bound of the inventory in the waste stream. However, in some cases, the assumed inventory can be exceeded. In these cases, DOE has a process to determine if it is acceptable to transfer the material to Tank 50.

NRC staff asked how the actual inventory being disposed is known if the inventory assumed in materials balance calculations represent an upper bound of the possible inventory. DOE contractor staff stated that as the actual Tank 50 sample data is made available, the inventory is updated to reflect the sample data, rather than the material balance information. DOE offered to provide a demonstration of the spreadsheet used for these calculations during the next onsite observation.

In this case, because the ARP/MCU sample data did not affect the inventory determination for the Saltstone Disposal Facility, NRC staff considers this follow-up action to be *closed*. However, NRC staff requests that it be informed when any major changes to the Salt Waste processes are made, such as exiting the ARP/MCU Management Control Plan, as these types of changes will affect the NRC's monitoring activities.

- *Follow-up Action: Anchor Bolt Penetrations in Vault 4*

    During the July 2010 onsite observation, NRC and DOE staffs discussed leakage from the vault caused by anchor bolts on the floors of cells 2A and 2B during the hydrotests (no waste involved). NRC staff raised a concern with the integrity of Vault 1 and 4 floors based on the presence of a similar drain system and anchor bolts. NRC staff suggested that direct evidence of leakage could be determined by horizontal soil cores under Vaults 1 and 4. DOE contractor staff described the progress that has been made towards addressing this follow-up action to date. DOE has visually inspected several anchor bolts locations in cells B and H of Vault 4, which are empty, and did not see any evidence of cracking on the vault floor surface. DOE also discussed historical, semi-annual monitoring well data that does not indicate that there have been releases from Vault 4. The potential effects of bolt penetrations will be mitigated in the FDCs, and the need for bolts (to anchor cable brace) will be eliminated in the new design.

    This follow-up action remains open pending the response to the above concern and the completion of the work DOE is performing on this follow-up action. The NRC staff will continue to review documentation regarding this follow-up action as it becomes available.

- *Follow-up Action: Impact of Scale on Core Sampling Methodology*

    During the July 2010 onsite observation, NRC and DOE staffs discussed proposed future saltstone core sampling techniques. Alternate methods were discussed and each had its strengths and weaknesses. DOE presented information on an in-situ sampling technique essentially using embedded pipes, for which they tested the force required to remove the sampling device. NRC expressed concern that the sampling device may allow less disruption of the sample, however the sampling device may change the in-situ conditions of the wasteform such that the sample is not representative. The NRC stated that when its contractor conducted experiments to test the properties of large-scale samples, scale effects were evident in the results. This highlights the importance of measuring properties of representative samples at appropriate scale.

    DOE is developing a formed-core sampling methodology to minimize the disruption to core samples that was discussed in the July 2010 onsite observation. NRC staff has commented that formed-core samples may not be representative of in-situ conditions, but it will continue to review core-sampling approaches and results.

    DOE will move forward with formed-core sampling technology. Operational considerations such as worker exposure and logistics will be considered in sampling plan. This follow-up action remains open pending the response to the completion of the work DOE is performing on the impact of scale on core sampling methodology.

Discussion Topic	NRC Question	Discussion Points
Cure Temperatures and Impact of Aluminate Concentration (October 2007 Observation)	The NRC onsite observation report for the October 2007 discussed the use of thermocouples within the vault and saltstone to monitor temperatures. At that point, the maximum observed temperature was ~50°C. What temperatures have been monitored since 2007 and what is the anticipated curing temperature based on the increase in aluminate concentration?	NRC staff inquired about the cure temperatures for saltstone grout as recent research has indicated its potential significance on the hydraulic properties of saltstone (WSRC-STI-2009-00419). A hydraulic conductivity of 8.6E-7 cm/s was measured for a saltstone grout simulant that was cured at 60°C, which is greater than the value assumed in the PA by more than a factor of 400. DOE stated that cure temperature profiles for saltstone are being compiled and will be considered in future testing. NRC staff discussed the importance of mimicking field conditions when practicable, including cure temperature and humidity.
Saltstone Fracturing (March 2009 Observation)	NRC staff noted that the saltstone fractures do not appear to be extensive, but that conclusion was hindered by lack of scale and a limited survey area and that DOE planned to do additional surveys in the future. Have there been any recent saltstone surveys in addition to Vault 4, Cell G?	During the March 2009 observation, participants watched a video survey that showed fractures on the surface of the saltstone grout in Cell G of Vault 4. The survey area was limited and DOE has since developed a video surveillance program to further evaluate fracturing of the saltstone surface. The video will be analyzed by DOE and evaluated with respect to the PA. NRC staff will review the video and analysis as they become available.
Hydrotest Results (April 2010 Observation)	What are the details of the recent hydrotests results for each cell (e.g., hydraulic head, test duration, observation procedures, observations)?	DOE stated that the hydrostatic testing of the new disposal cells, following the design changes showed no evidence of leaking. The follow-up testing consisted of a modified version of the earlier hydrotest. The new test consisted of a 12-foot head differential for 132 hours. The NRC staff considers this item to be closed.
Vault 4 Floor (July 2010 Observation)	Vault 4 Floor - NRC staff asked that DOE look into characterizing the Vault 4 floor (or the material under the floor) to see if the floor in Vault 4 had cracked. Has there been any progress on this topic?	The response to this email question was discussed in the Follow-up Action: Anchor Bolt Penetrations in Vault 4 section. This follow-up action remains open pending the response to the above question and the completion of the work DOE is performing on this follow-up action.

13

2.4.3 Conclusions and Follow-up Actions:

No new issues or concerns were identified during the technical discussion; however, multiple follow-up actions were identified during the discussion. Below is a list of items discussed during this portion of the observation.

Performance Assessment/Research Activity - NRC staff would like to know the results of the current research activities regarding the current and future SDF PA maintenance activities. This is not a follow-up action; however, the staff maintains an interest in PA maintenance activities and will continue discussions with the DOE leading up to its upcoming revision to the 2007 NRC monitoring plan for the Saltstone Disposal Facility.

Open Issues 2007-1 and 2007-2 - NRC staff indicated that DOE's plans to address the open issues sound reasonable and will review the results of DOE's planned efforts. *Currently, the two issues remain open.*

Follow-up Action: Disposal Unit 2 Water Tightness Test Quality Assurance Records - DOE stated they would provide the NRC staff documentation of the design changes and hydrotesting results of the new disposal cells. NRC will review these documents when provided. *This follow-up action remains open.*

Follow-up Action: Radiological Composition of Inadvertent Transfer Material - Based on the information provided to the NRC in SRR-WSE-2010-00186 (ML111780337) provided to the NRC on October 26, 2010, *the NRC staff considers this action item to be closed.*

Follow-up Action: Status of ARP/MCU Management Control Plan - The NRC would like DOE to clarify how the sampling data obtained under the ARP/MCU management control plan is used and how the inventory of radionuclides sent to Tank 50 from ARP/MCU is determined. *This is a new follow-up action from this observation.*

In response to the DOE concern that tracking of long-term items as follow-up actions might not be the most efficient mechanism, NRC staff stated that the new monitoring plan will contain a list of major changes to the salt waste disposal process that they would like to be made aware of, if and when they occur.

DOE offered to provide a demonstration of the spreadsheet used for these inventory-updating calculations during the next onsite observation. This is not a follow-up action; however, the NRC would like to observe this demonstration in the future. The NRC makes note that this will be a future observation activity.

Because the ARP/MCU sample data does not affect the inventory determination for the Saltstone Disposal Facility, *NRC staff considers this follow-up action, which was created during this observation, to be closed.*

14

Follow-up Action: Anchor Bolt Penetrations in Vault 4 - Because of the pending response to the NRC concern regarding the integrity of Vault 1 and 4 floors based on the presence of a similar drain system and anchor bolts and the pending completion of the work DOE is performing on the impacts of anchor bolt penetrations, this follow-up action remains open.

Follow-up Action: Impact of Scale on Core Sampling Methodology - NRC staff will continue to review core-sampling approaches and results, and DOE will move forward with formed-core sampling technology with consideration of the discussed exposure and logistical techniques when developing the sampling plan.

Cure Temperatures and Impact of Aluminate Concentration - NRC staff will review the cure temperature profiles for saltstone when DOE compiles them following future testing.

Saltstone Fracturing - NRC will review saltstone surface fracturing surveillance captured by DOE's recently developed video surveillance program as they become available.

Hydrotest Results - DOE stated that the hydrostatic testing of the new disposal cells, following the design changes showed no evidence of leaking. The follow-up testing consisted of a modified version of the earlier hydrotest. The new test consisted of a 12-foot head differential for 132 hours. The NRC staff considers this follow-up action to be closed.

3.0   OVERALL CONCLUSIONS AND FOLLOW-UP ACTIONS:

3.1   Technical Discussion – Saltstone Radionuclide Inventory:

Of the three questions discussed during this portion of the observation, only question (2) remains unanswered. Question (2) will be addressed in DOE's response to the NRC staff's second RAI (RAI-2009-02). No additional issues or concerns were identified during the technical discussion on the radionuclide inventory of Vault 4 apart from DOE's continued effort to respond to RAI-2009-02. The NRC continues to have reasonable assurance that the 10 CFR Part 61 performance objectives can be met provided that key assumptions made in the waste determination prove to be correct.

3.2   Technical Discussion – New Research on Long-Term Testing Waste Oxidation and Technetium Release:

Because of the NRC staff's concerns discussed above, Open Issue 2009-1 remains open. No additional issues or concerns were identified during the technical discussion regarding technetium release and oxidation in the saltstone wasteform. The NRC continues to have reasonable assurance that the 10 CFR Part 61 performance objectives can be met provided that key assumptions made in the waste determination prove to be correct.

3.3   Discussion of Disposal Unit 2 Construction:

The NRC staff will continue to monitor the construction of the new disposal cells and will continue to monitor the cells when they are put into operation.

3.4   Follow-up Discussion – Topics from Previous Observations:

The table below summarizes the status of each of the discussion topics. Each topic is classified as being open, closed, or a future topic for discussion. The following terms are used to classify the topics.

Remains Open: The NRC is still awaiting action on the part of DOE, or results from a recent action taken by DOE. Further discussion will need to take place before the NRC can close the topic.

Topic Closed: The specific inquiry posed by the NRC has been fully responded to by DOE.

Future Consideration: The specific inquiry posed by the NRC has been discussed and DOE has stated a path forward that seems acceptable to the NRC. The item is not open because the DOE plans to address the topic. The item is not closed because the NRC is interested in the results of the analysis being performed by DOE.

Discussion Topic	Remains Open	Topic Closed	Future Consideration
PA/Research Activity			X
Open Issues 2007-1 and 2007-2	X		
Follow-up Action: Disposal Unit 2 Water Tightness Test Quality Assurance Records	X		
Follow-up Action: Radiological Composition of Inadvertent Transfer Material		X	
Follow-up Action: Status of ARP/MCU Management Control Plan		X	X
Follow-up Action: Anchor Bolt Penetrations in Vault 4	X		
Follow-up Action: Impact of Scale on Core Sampling Methodology			X
Cure Temperatures and Impact of Aluminate Concentration			X
Saltstone Fracturing			X
Hydrotest Results		X	
Vault 4 Floor	X		

4.0   PARTICIPANTS:

U.S. NRC	U.S. DOE-SR	SRR
George Alexander	Sherri Ross	Ginger Dickert
Christepher McKenney	Patricia Suggs	F. Malcolm Smith
Andrew Persinko		Aaron Staub
Karen Pinkston		Steve Thomas
Christianne Ridge		Rebecca Freeman
James Shaffner (*for Nishka Devaser*)		

U.S. DOE
Linda Suttora

5.0   REFERENCES:

U.S. Department of Energy (DOE). CBU-PIT-2005-0146, "Saltstone Performance Objective Demonstration Document." Westinghouse Savannah River Company. June 2005.

——. DOE-WD-2005-001, "Basis for Section 3116 Determination Salt Waste Disposal at the Savannah River Site." Washington, DC: DOE. January 2006.

——. SRR-CWDA-2009-00017, Performance Assessment for the Saltstone Disposal Facility at the Savannah River Site, Savannah River Site, Aiken, SC, October 29, 2009 (ADAMS Accession No. ML101590008). October 2009.

U.S. Nuclear Regulatory Commission (NRC). "Technical Evaluation Report for Draft Waste Determination for Salt Waste Disposal." Letter from L. Camper to C. Anderson, DOE. December 28, 2005. (Agencywide Documents Access and Management System (ADAMS) Accession No. ML053010225).

——. "U.S. Nuclear Regulatory Commission Plan for Monitoring the U.S. Department of Energy Salt Waste Disposal at the Savannah River Site in Accordance with the National Defense Authorization Act for Fiscal Year 2005." Washington, DC: NRC. May 3, 2007. (ADAMS Accession No. ML070730363).

——. "U.S. Nuclear Regulatory Commission March 24-28, 2008 Onsite Observation Report for the Savannah River Site Saltstone Facility." June 5, 2008 (ADAMS Accession No. ML081290367).

——. "U.S. Nuclear Regulatory Commission March 25-26, 2009 Onsite Observation Report for the Savannah River Site Saltstone Facility." May 22, 2009 (ADAMS Accession No. ML091320439).

——. "U.S. Nuclear Regulatory Commission Request for Additional Information on the 2009 Performance Assessment for the Saltstone Disposal Facility at the Savannah River Site." March 31, 2010a (ADAMS Accession Number ML100820097).

——. "U.S. Nuclear Regulatory Commission April 19, 2009 Onsite Observation Report for the Savannah River Site Saltstone Facility." July 7, 2010b (ADAMS Accession No. ML101460044).

NRC FORM 335 (12-2010) NRCMD 3.7	U.S. NUCLEAR REGULATORY COMMISSION **BIBLIOGRAPHIC DATA SHEET** *(See instructions on the reverse)*	1. REPORT NUMBER (Assigned by NRC, Add Vol., Supp., Rev., and Addendum Numbers, if any.) NUREG-1911, Revision 4

2. TITLE AND SUBTITLE		3. DATE REPORT PUBLISHED	
NRC Periodic Compliance Monitoring Report for U.S. Department of Energy Non-High-Level-Waste Disposal Actions: Annual Report for Calendar Year 2011		MONTH August	YEAR 2012
		4. FIN OR GRANT NUMBER	
5. AUTHOR(S)		6. TYPE OF REPORT Annual	
		7. PERIOD COVERED (Inclusive Dates) 01/01/2011 - 12/31/2011	

8. PERFORMING ORGANIZATION - NAME AND ADDRESS (If NRC, provide Division, Office or Region, U. S. Nuclear Regulatory Commission, and mailing address; if contractor, provide name and mailing address.)

Division of Waste Management and Environmental Protection
Office of Federal and State Materials and Environmental Management Programs
U.S. Nuclear Regulatory Commission
Washington, DC 20555-0001

9. SPONSORING ORGANIZATION - NAME AND ADDRESS (If NRC, type "Same as above", if contractor, provide NRC Division, Office or Region, U. S. Nuclear Regulatory Commission, and mailing address.)

Same as NRC address above

10. SUPPLEMENTARY NOTES
PROJ0734, PROJ0735

11. ABSTRACT (200 words or less)

This is the U.S. Nuclear Regulatory Commission (NRC) staff's report of its monitoring of U.S. Department of Energy (DOE) non-high-level waste disposal actions in calendar year 2011, in accordance with Section 3116(b) of the Ronald W. Reagan National Defense Authorization Act for Fiscal Year 2005 (the NDAA). Section 3116 of the NDAA requires that DOE consult with the NRC on its non-high-level waste determinations and plans and that the NRC, in coordination with the covered States of South Carolina and Idaho, monitor disposal actions that DOE takes to assess compliance with NRC regulations in Title 10 of the Code of Federal Regulations (10 CFR) Part 61, "Licensing Requirements for Land Disposal of Radioactive Waste," Subpart C, "Performance Objectives." The NRC has prepared this report in accordance with NUREG 1854, "NRC Staff Guidance for Activities Related to U.S. Department of Energy Waste Determinations," issued August 2007.

12. KEY WORDS/DESCRIPTORS (List words or phrases that will assist researchers in locating the report.)	13. AVAILABILITY STATEMENT unlimited
Incidental waste, High-level waste tanks, Waste determinations, WIR, Waste incidental to reprocessing, Savannah River Site, Idaho National Laboratory, Hanford, C-Tank Farm, F-Tank Farm, Saltstone Disposal Facility, Section 3116 of the NDAA, Tank Farm Facility	14. SECURITY CLASSIFICATION (This Page) unclassified (This Report) unclassified
	15. NUMBER OF PAGES
	16. PRICE

NRC FORM 335 (12-2010)

Printed
on recycled
paper

Federal Recycling Program

NUREG-1911, Rev. 4

NRC Periodic Compliance Monitoring Report for
U.S. Department of Energy Non-High-Level Waste Disposal Actions

August 2012

www.ingramcontent.com/pod-product-compliance
Lightning Source LLC
Chambersburg PA
CBHW080258180526
45167CB00006B/2579